PHOTONIC CRYSTALS

Edited by **Alireza Bananej**

Photonic Crystals
http://dx.doi.org/10.5772/58717
Edited by Alireza Bananej

Contributors

Igor Lyubchanskii, Nataliya Dadoenkova, YoungPak Lee, Theo Rasing, Yulia Dadoenkova, Andrey Zabolotin, Hanben Niu, Chun-Feng Lai, Ahmad Naseri Taheri, Hassan Kaatuzian, Alagappan Gandhi, Alessandro Settimi

Published by InTech

Janeza Trdine 9, 51000 Rijeka, Croatia

Publishing Process Manager Iva Lipovic

Technical Editor InTech DTP team

Cover InTech Design team

First published April, 2015

Printed in Croatia

Additional hard copies can be obtained from orders@intechopen.com

Photonic Crystals, Edited by Alireza Bananej

p. cm.

ISBN 978-953-51-2121-3

For EU product safety concerns:
IN TECH d.o.o., Prolaz Marije Krucifikse Kozulić 3, 51000 Rijeka, Croatia,
info@intechopen.com or visit our website at intechopen.com.

Meet the editor

Dr. Alireza Bananej received his B.Sc degree in applied physics in 1989. Then he joined Laser Research Center of NSTRI and worked in the field of optical coating technology. In 1996 he received his M.Sc degree from Amirkabir University of Technology in applied physics. In 2000 he joined the physics department, nonlinear photonics lab of Harbin Institute of Technology (HIT) in China and worked in the field of low power and controllable all optical switching. He received his Ph.D degree from HIT in 2004 and returned to Laser Research Center of NSTRI. Since 2004 he has been studying fiber laser, high power and fast optical coating and application of photonic crystals.

Contents

Preface

As a reason of information explosion, electronic communication networks are not sufficient for high bit rate data transmission. This problem has been solved by optical networks which caused the birth of a new area of technology, photonics. In photonic circuits photons play the dominant role and they transfer the optical data. With the growth of the photonics technology, a new area started to grow as photonic crystals which now play an important role in designing and manufacturing compact photonic devices.

Photonic crystals are structures with alternative dielectric constant in one, two or three dimensions which are called one, two or three dimensional photonic crystals. As a reason of alternative dielectric constant, by solving Maxwell's equations photonic band gap can appear. By using the properties of photonic band gap, many interesting phenomena such as slow light generation, dispersion engineering in a compact and low size device can be achieved.

Alireza Bananej
Laser and Optics Research School,
NSTRI, IRAN

One-Dimensional Photonic Crystals With the Superconducting Defects

N. N. Dadoenkova, A. E. Zabolotin,
Yu. S. Dadoenkova, I. L. Lyubchanskii, Y.P. Lee and
Th. Rasing

Additional information is available at the end of the chapter

http://dx.doi.org/10.5772/59971

1. Introduction

During the last 27 years, starting from pioneering papers by E. Yablonovich [1] and S. John [2], photonic crystals (PCs) or photonic band gap (PBG) materials are the objects of intensive theoretical and experimental research, because of promising applications in modern photonics and related areas [3–6]. Usually, PCs are artificial one-, two- or three-dimensional (1D, 2D, and 3D) periodic structures with periods which are comparable with the wavelengths of electromagnetic waves (EMWs) and constructed of materials with different refractive indices [3–6]. The PBGs are forbidden regions in the dispersion law and in transmittivity spectra, where EMWs with selected frequencies cannot propagate through the PC. A lot of attention has been directed to the study of conventional PCs on the base of dielectrics, semiconductors, and normal metals as well as on the base of functional materials like magnetics [7, 8], ferroelectrics [9, 10] and liquid crystals [6, 11, 12] which can be controlled by external magnetic or electric fields. Another interesting group of PCs are those with superconducting (SC) constituents, so-called superconducting PCs. Scientific activity in different aspects of SC PC is described in several original and review papers by S. Savel'ev et al. [13 – 15], and in the review paper by S. Anlage [16]. The majority of papers about SC PC is devoted to regular periodic photonic structures. But, if the periodicity of a PC is destroyed by the introduction of a so-called defect unit, for example a defect layer into a 1D PC, the transmittivity spectra are drastically changed. In that case, defect modes (DMs) emerge inside the PBG because of localization of EMWs around the defect unit.

These DMs can be observed as very narrow peaks (in comparison with the PBG widths) with sufficiently large transmittivity. If such defects are made of functional materials which are sensitive to the action of various external factors like electric or magnetic fields or temperature, it is possible to control the transmittivity of PCs at the DM frequencies.

In our papers [17, 18] we investigated the influence of a complex defect layer constructed from dielectric and SC sublayers on the transmittivity of PCs for different thicknesses of the SC constituent and temperatures for normal [17] and oblique [18] incidence of EMWs on the PC.

In [19] it was shown that a thin SC layer on the top of a dielectric PC leads to selective transmittivity for different polarizations of incident EMWs. The influence of a complex defect layer (composed of SC and dielectric sublayers) as a spacer between two different PCs was studied in [20]. The localized modes in a metamaterial-dielectric PC with a dielectric-SC pair defect were investigated in [21]. The behavior of a DM inside a PBG depending on position (SC-dielectric and dielectric-SC) in a 1D PC was analyzed in [22]. In a recent paper [23], the normal and oblique incidence of an EMW on 1D dielectric PC with a thin SC defect layer was theoretically investigated and the obtained results show qualitative agreement with our calculations [17, 18]. An interesting situation takes place when two (or more) defects are introduced into a PC. In the case of two defects two DMs appear inside PBG and their mutual positions strongly depend on the distance between the defect units. The linear and nonlinear optical, as well as, magneto-optical properties of 1D PCs with two magnetic defects were theoretically studied in a few papers [24 – 27]. In these papers some interesting peculiarities of the spectra of DMs inside the PBGs, like the possibilities to obtain relatively wide peaks formed by two DMs, were obtained and discussed. One expect that in PCs with two complex defects composed of SC and dielectric sublayers one can obtain new peculiarities as well.

In this chapter, we theoretically investigate the transmittivity, reflectivity and absorptance of 1D PCs with one and two complex defects containing SC constituents. Two cases of asymmetric positions of complex SC-containing defects (when both SC-sublayers are located on the right-hand, or on the left-hand sides of the dielectric defect sublayers inside the photonic structure) are investigated, as well as the symmetric ones. We also analyze the dependencies of the optical properties of such PCs as functions of SC-sublayer thicknesses and temperature. The chapter is organized as follows. In Sec. 2 we describe the aforementioned photonic structure with one complex defect layer with an analytical approach. In Sec. 3 the numerical calculations of the transmittivity spectra for the photonic structure two superconducting defect layers are presented. In Sec. 4, conclusions, we summarize the obtained results.

2. Photonic crystal with one defect

Let us consider finite 1D periodic photonic structure composed of two finite size PCs of structure $(AB)^5$ with the period $D=d_1+d_2$, where the layer A is the strontium titanate $SrTiO_3$ with thickness d_1, the layer B is the aluminum oxide Al_2O_3 with thickness d_2, and the complex defect layer placed between these PCs, as shown in Fig. 1. We assume the medium surrounding the photonic structure to be a vacuum. The layers of PC are located in the xy-plane and the z-axis to be normal to the interfaces. The defect layer consists of SC sublayer $YBa_2Cu_3O_7$ of thickness d_s and dielectric sublayer $SrTiO_3$ of thickness d_{1d}, respectively. The selection of these dielectrics is due to the fact that both these materials $SrTiO_3$ and Al_2O_3 are widely used as substrates for $YBa_2Cu_3O_7$ SC films.

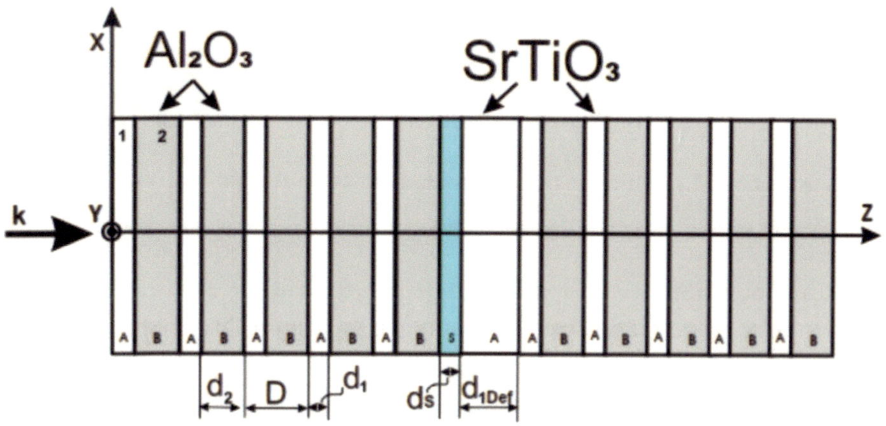

Figure 1. Schematic of photonic structure which is containing of two finite-size PCs with complex SC-dielectric defect: $(AB)^5(Sc)(defA)(AB)^5$.

For the SC layers, we consider the case when the crystallographic *a*-axis (electrically "light" axis) and *c*-axis (electrically "heavy" axis) of $YBa_2Cu_3O_7$ coincide with the *x*- and *y*-axes, respectively. In our study we used the frequency- and temperature-dependent dielectric permittivity tensor for $YBa_2Cu_3O_7$ with nonzero diagonal components $\varepsilon_{xx}(\omega, T)=\varepsilon_{zz}(\omega, T)$ and $\varepsilon_{yy}(\omega, T)$, introduced in Ref. [19] on the base of a generalized two-fluid model in the following form:

$$\varepsilon_{s,v}(\omega,T) = 1 + \frac{4\pi i}{\omega}\sigma_{s,v}(\omega,T), \quad v = xx,\ yy. \tag{1}$$

Here the dynamic electrical conductivity tensor $\sigma_{s,v}(\omega, T)$ is defined as

$$\sigma_{s,v}(\omega,T) = f_+(\omega,T)\sigma^+_{s,v}(\omega,T) + f_-(\omega,T)\sigma^-_{s,v}(\omega,T), \tag{2}$$

where $\sigma^{+}_{s,v}(\omega, T)$ are determined as follows:

$$\sigma^-_{s,v}(\omega,T) = \frac{T_c}{\Lambda^{(0)}_{s,v}(\alpha T - i\omega T_c)}, \quad \sigma^+_{s,v}(\omega,T) = \frac{i(T^4 - T_c^4)}{\omega\Lambda^{(0)}_{s,v}T_c^4} + \left[\frac{T}{T_c}\right]^4 \sigma^-_{s,v}(\omega,T), \tag{3}$$

where T_c and $f_{\pm}(\omega, T)$ are the critical temperature and the electron distribution functions, respectively,

$$f_{\pm}(\omega,T)=\left(1+\exp\left(\frac{\pm\hbar\left(\omega-\omega_s(T)\right)}{k_B T}\right)\right)^{-1}, \tag{4}$$

where k_B and $\hbar$ are the Boltzmann's and Planck's constants, respectively. In Eq. (4) the pair braking frequency ω_s is determined via the temperature-dependent superconductor half-gap energy $\Delta(T)$ as $\hbar\omega_s=\Delta(T)$, which is approximated as $\Delta(T)\cong\Delta_0\left(1-\frac{T^3}{T_c^3}\right)^{1/4}$ (see for details Ref. [19]).

In the optical and near infra-red regimes the electrodynamical properties of $YBa_2Cu_3O_7$ can be described by a dielectric permittivity only. The magnetic permeability of $YBa_2Cu_3O_7$ is assumed to be $\mu_s=1$.

It should be noted that SC sublayer exhibits strongly pronounced anisotropy of optical properties. Both components of the permittivity tensor $\varepsilon_{s,xx}$ and $\varepsilon_{s,yy}$ are complex values, moreover, in considered frequency region (first PGB) both real and imaginary parts of $\varepsilon_{s,xx}$ are about two order of magnitude larger then corresponding ones for $\varepsilon_{s,yy}$. As $\varepsilon_{s,yy}$ and $\varepsilon_{s,xx}$ are responsible for TE- and TM- polarized modes, respectively, it means that decaying of TM-polarized mode inside SC sublayer is much stronger then of TE-polarized one. This difference in values of $\varepsilon_{s,xx}$ and $\varepsilon_{s,yy}$ leads to drastic contrast in behavior TE- and TM- modes with variation of the SC sublayer thickness d_s.

We investigate the case of the normal incidence of light on the right hand surface of the PCs. We assume the incident light to be linearly polarized: *x*- or *y*- polarizations (the EMWs with electric field vector **E** vibrations along the *x*- or *y*-axis, respectively).

We calculate the transmittivity and reflectivity spectra of the 1D PC using the four-dimensional transfer matrix method by Berreman [29]. The details of the method are given in our previous papers [17, 18, 30].

In the case of non-SC defect layer (single dielectric layer with thickness $d_{def}=d_1$) symmetrically embedded between two identical PCs, the dependence of transmittivity for both TE- and TM-modes in 1D PC is characterized by the presence of defect mode in the center of PBGs, as depicted on Fig. 2.

As shown in Ref. [19], a thin SC layer, deposited at one side of dielectric PC, leads to the decrease of transmittivity for the EMW, polarized along the *x*-axis (TM-mode), whereas TE-polarized EMW propagates through this photonic structure practically without losses.

Below we analyze numerically both intensity and position of the defect mode inside the first PBG as a function of thickness of SC sublayer for the system under consideration (see

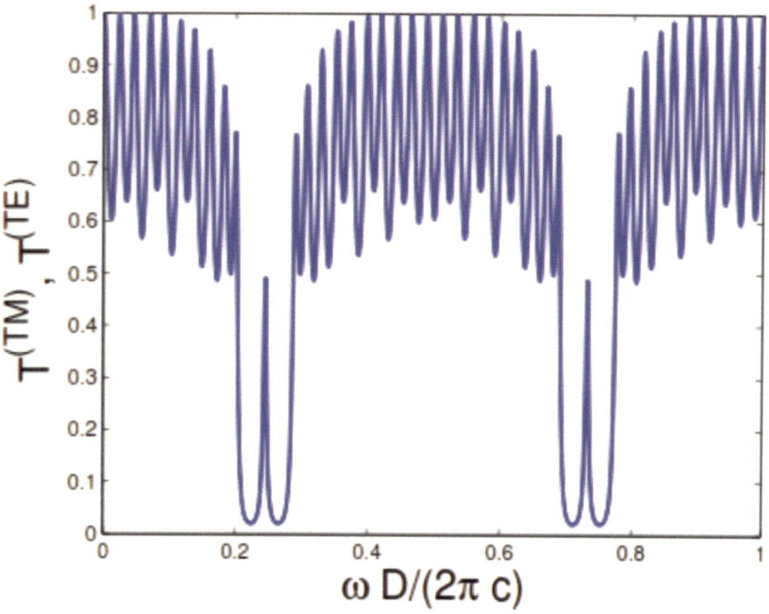

Figure 2. First two PBGs with defect modes located in the centers of forbidden zones. The transmittivities for TE- and TM- modes coincide in the case of defect without SC sublayer.

Fig. 1). From the numerical calculations we obtained, that position and intensity of the defect mode are changing significantly for TM- mode. The corresponding results are presented on Figs. 3 – 6.

We calculated the intensities and positions of both TE- and TM-polarized defect modes inside the first PBG as a function of SC sublayer thickness for different temperatures: T=4.2 K, T=77 K (below temperature of SC transition) and T=90 K (at temperature of SC transition). The refractive indices for PC's constituents are equal to n_1=2.437, n_2=1.767 for $SrTiO_3$ and Al_2O_3, respectively [28] and the corresponding thicknesses are d_1=0.42 D, d_2=0.58 D, where PC's period is chosen to be D=5 μm. We varied the SC sublayer thickness in the region between 0 and 70 nm, keeping the thickness of dielectric defect sublayer to be d_{1d}=0.42 D=2.1 μm.

For an estimation of the frequency and temperature dependence of the dielectric tensor of $YBa_2Cu_3O_7$ $\varepsilon_{s,v}$ $(v=xx,\ yy,\ zz)$ (see Eqs. (14) – (16) in paper [18]) we used the following parameters: the scattering rate component α=1.4•10^{13} s^{-1}, the London tensor components at absolute zero $\Lambda_{s,x}^{(0)}=2.1\cdot10^{-30}$ s^2 and $\Lambda_{s,y}^{(0)}=1.9\cdot10^{-28}$ s^2, as well as the half-gap energy at absolute zero Δ_0= 3.0•10^{-2} eV [19].

The results of our calculations for TE-mode are presented in Figs. 3 – 5 and for TM-mode are given in Fig. 6. One can see that the position of TM-polarized defect mode strongly depends on the SC sublayer thickness: the peak of the defect mode shifts to the right edge of the PBG with the increase of d_s.

From Fig. 3 one can see, that at the temperature *T=4.2 K* the defect mode intensity is monotonically increasing with growth of SC sublayer thickness, and the corresponding defect peak approaches the right edge of the PBG.

At higher temperatures (*T=77 K* and *T=90 K*) the intensity of the defect mode fist is going down to with growth of d_s, and then goes up and merge with the right PBG edge.

In contrast of TM-mode, the defect mode of TE-polarization almost does not change its position under growth of temperature and SC sublayer thickness.

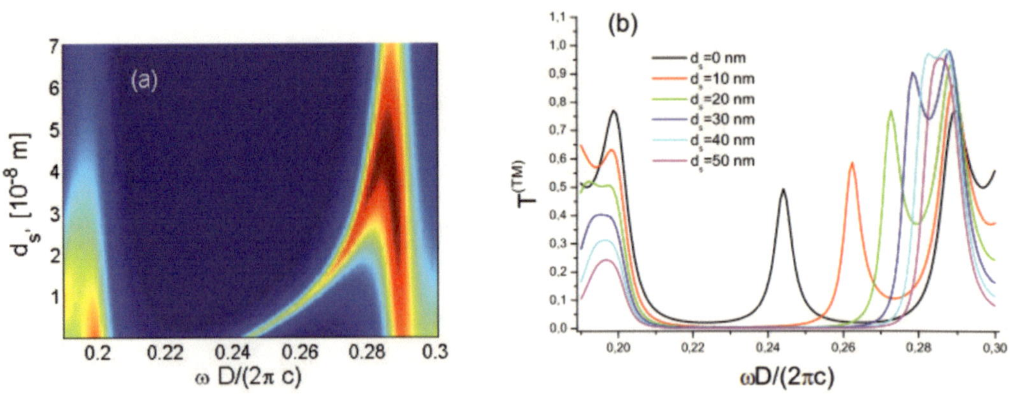

Figure 3. The intensity and position of the TM-polarized defect mode inside the first PBG as function of SC sublayer thicknesses d_s and normalized frequency $\tilde{\omega}=\omega D/(2\pi c)$ at the temperature *T=4.2 K*: (a) the top view; (b): the profiles of the defect mode for different SC sublayer thicknesses.

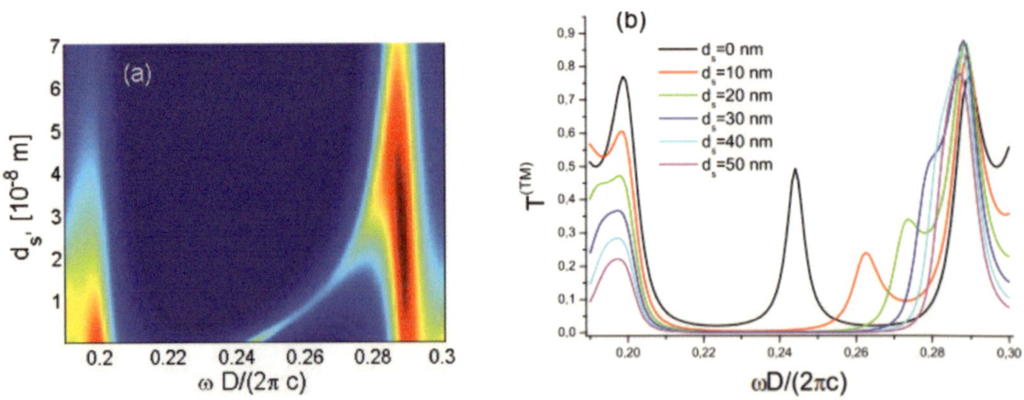

Figure 4. The intensity and position of the TM-polarized defect mode inside the first PBG as function of SC sublayer thicknesses d_s and normalized frequency $\tilde{\omega}=\omega D/(2\pi c)$ at the temperature *T=77 K*: (a) the top view; (b): the profiles of the defect mode for different SC sublayer thicknesses.

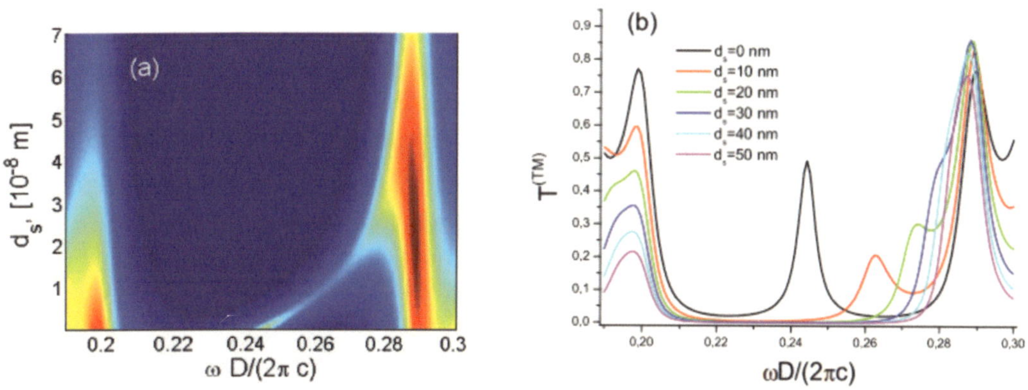

Figure 5. The intensity and position of the TM-polarized defect mode inside the first PBG as function of SC sublayer thicknesses d_s and normalized frequency $\widetilde{\omega}=\omega D/(2\pi c)$ at the temperature *T=90 K*: (a) the top view; (b): the profiles of the defect mode for different SC sublayer thicknesses.

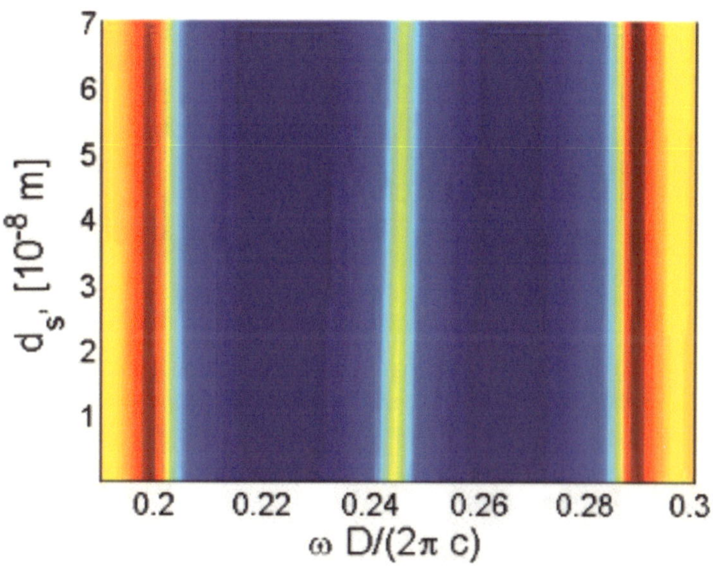

Figure 6. The top view of intensity and position of the TE-polarized defect mode inside the first PBG as function of SC sublayer thicknesses d_s and normalized frequency $\widetilde{\omega}=\omega D/(2\pi c)$.

3. Photonic crystal with two defects

Let us consider a finite size 1D PC of the structure $(BA)^N\ Def_2(BA)^M\ Def_1(BA)^N$ consisting of $2N+M$ regular unit cells (BA) with two combined defect layers Def_1 and Def_2 embedded into the PC symmetrically with respect to the PC's edges, as shown in Fig. 1(a). The unit cell with period

$D=d_1+d_2$ of the PC is formed by two dielectric layers: the layer A of strontium titanate $SrTiO_3$ with thickness d_1, and the layer B of aluminum oxide Al_2O_3 with thickness d_2. All the layers are located in the xy-plane and the z-axis is perpendicular to the interfaces. Both complex defects Def_1 and Def_2 consist of a $SrTiO_3$ sublayer A_{def} of thickness d_{1d} and a SC defect sublayer $YBa_2Cu_3O_7$ of thickness d_s. The selection of these dielectrics is due to the fact that both $SrTiO_3$ and Al_2O_3 are widely used as substrates for $YBa_2Cu_3O_7$. The medium surrounding the PC is vacuum.

We distinguish the two types of the complex defect layers, depending on the positions of the SC sublayers with respect to the dielectric defect ones: the right-handed (RH) defects $Def_R=A_{def}$ SC and left-handed (LH) defects $Def_L=SC\ A_{def}$. In Figs. 7(a) – 7(d) we depicted the schematic of four possible geometries of the PCs with two combined SC defects of RH and LH types. The Figs. 7(a) and 7(b) demonstrate the PCs with two RH and two LH defects, which further will be referred to as RH – RH and LH – LH geometries, respectively. The PCs with the SC defects of different types: RH – LH and LH – RH are shown in Figs. 7(c) and 7(d), respectively.

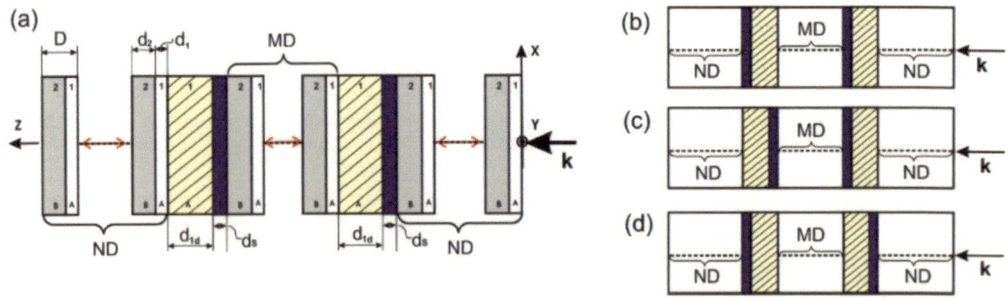

Figure 7. Schematic of finite two-component 1D PCs with two combined SC defect layers: RH – RH (a), LH – LH (b), RH – LH (c), and LH – RH (d) geometries.

Below we present the numerical calculations of the transmittivity spectra of the PCs under consideration as function of the SC defect sublayer thickness d_s, as well as of the dielectric defect sublayer thickness d_1. Also, we numerically investigate the temperature dependence of the corresponding spectra. We restrict our considerations to the frequency range within the first PBG and its vicinity. For the numerical calculations presented below we chose the same parameters of the PC with unit cell numbers of $N=5$ and $M=2$, as in Sec. 2.

3.1. PC with two RH defects

First, we start our investigation with a PC containing two RH defects (see Fig. 7(a)). In Figs. 8(a) and 8(b) we present the top view of the transmittivities $T^{(x)}$ for the x-polarized EMWs vs the normalized frequency $\tilde{\omega}$ (within the range of the first PBG and its vicinity) and the defect SC sublayer thickness d_s for the liquid helium and liquid nitrogen temperatures: $T=4.2\ K$ and $T=77\ K$, respectively. The SC sublayer thickness d_s varies within the range $(0 \div 70)$ nm, while the dielectric defect layer thickness is fixed to be $d_{1d}=d_1=2.1\ \mu$m. The presence of two combined defect layers in the PC leads to the appearance of two narrow DMs inside the PBG (see Figs.

8(a), 8(b)) which are referred to as low-frequency (LF) and high-frequency (HF) DMs. In Fig. 8(c) we plotted the transmittivity spectra for the cases of different SC sublayer thicknesses: d_s=0, 10, 20, 30, 40 and 50 nm, calculated for the temperature T=4.2 K.

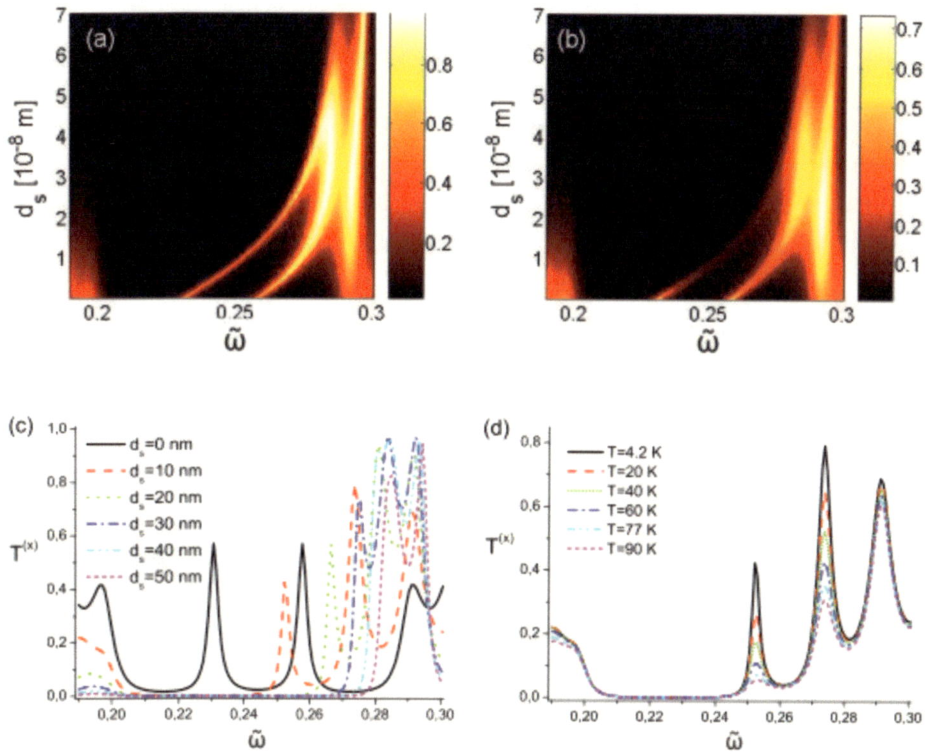

Figure 8. Two RH SC defects. Top view of the transmittivity $T^{(x)}$ vs normalized frequency $\tilde{\omega}$ and the defect SC sublayer thickness d_s for T=4.2 K (a) and T=77 K (b); the transmittivity spectra for different values of d_s calculated for T=4.2 K (c); and the temperature evolution of the transmittivity spectrum for d_s=10 nm (d). The results are obtained for the case of the dielectric defect layer with d_{1d}=d_1=2.1 μm.

We calculated the frequencies of the PBG edges for the PC with pure dielectric defects (when d_s=0): the LF PBG edge is $\tilde{\omega}_{LF}^{PBG}$ =0.19637 (which is equivalent to ω_{LF}^{PBG} =73.98 THz or λ_{LF}^{PBG} =25.46 μm and the HF PBG edge is $\tilde{\omega}_{HF}^{PBG}$ =0.29107 (which is equivalent to ω_{HF}^{PBG} =109.65 THz or λ_{HF}^{PBG} =17.18 μm). According to this estimation the first PBG's width is $\Delta\omega_{PBG}$ =35.68 THz, which corresponds to $\Delta\lambda_{PBG}$ =8.28 μm. In the case of pure dielectric defects (d_s=0 nm) the DMs are localized inside the PBG symmetrically with respect to the PBG edges at the normalized frequencies $\tilde{\omega}_{LF}^{peak}$ =0.23059 (which is equivalent to ω_{LF}^{peak} =86.87 THz or λ_{LF}^{peak} =21.68 μm) and $\tilde{\omega}_{HF}^{peak}$ =0.25765 ω_{HF}^{peak} =97.06 THz or λ_{HF}^{peak} =19.41 μm) and their magnitudes practically coincide: $T_{LF}^{(x)} \approx T_{HF}^{(x)} \approx$ 0.57 as shown by the black solid line in Fig. 8(c). The presence of the SC sublayer breaks the symmetry of the position and values of the DM's peaks. The increase of d_s leads to an essential shift of the DMs to the HF PBG edge up to their mergence near the HF PBG edge for $d_s \approx 40$ nm, as one can see in Figs. 2(a), 2(b) and 2(c). For example, for d_s=10 nm, the LF and HF peaks positions become $\tilde{\omega}_{LF}^{peak}$ =0.25207 (which is equivalent to ω_{LF}^{peak} =94.96 THz or λ_{LF}^{peak} =19.83 μm) and $\tilde{\omega}_{HF}^{peak}$ =0.25765 (ω_{HF}^{peak} =103.06 THz or λ_{HF}^{peak} =18.28 μm), respectively. The frequency shifts of the LF and HF DMs are $\Delta\tilde{\omega}_{LF}$ =0.02148 (which is equivalent to $\Delta\omega_{LF}^{peak}$ =8.09 THz or $\Delta\lambda_{LF}^{peak}$ =2.27 μm) and $\Delta\tilde{\omega}_{HF}$ =0.01591 ($\Delta\omega_{HF}^{peak}$ =6.02 THz or $\Delta\lambda_{HF}^{peak}$ =1.55 μm) which are about 27.4% and 18.7% of the PBG width at d_s=0 nm, respectively. The distance between the DMs decreases considerably with the increase of the SC sublayer thickness: from $\Delta\tilde{\omega}$ =0.02706 ($\Delta\omega^{peak}$ =10.19 THz or $\Delta\lambda^{peak}$ =2.27 μm) for d_s=0 to $\Delta\tilde{\omega}$ =0.02149 ($\Delta\omega^{peak}$ =8.01 THz or $\Delta\lambda^{peak}$ =1.55 μm) for d_s=10 nm and to $\Delta\tilde{\omega}$ =0.01432 ($\Delta\omega^{peak}$ =5.39 THz or $\Delta\lambda^{peak}$ =0.96 μm) for d_s=20 nm, to $\Delta\tilde{\omega}$ =8•10^{-3} ($\Delta\omega^{peak}$ =2.99 THz or $\Delta\lambda^{peak}$ =0.51 μm) for d_s=30 nm.

The SC sublayer thickness increase also leads to an abrupt reduction of the transmittivity value at the LF PBG edge. The HF PBG edge position changes as well, slightly moving to higher frequencies with increasing of d_s and reaching the maximum value $T^{(x)}$≈0.96 for d_s=40 nm. From Fig. 8(a) one can see that for the SC defect layer thickness d_s=40 nm both DMs merge. The amplitudes of the peaks vary as well: the transmittivity at the HF DM grows with the increase of d_s, while the LF peak intensity first goes down to $T_{\min}^{(x)} \approx 0.42$ (at $\tilde{\omega}$ ≈0.252) for d_s=10 nm, while after that, its amplitude also increases with further increase of d_s.

The DMs intensities change considerably with increasing temperature. Comparing Figs. 8(a) and 8(b), one can see that the influence of temperature on the PBG spectra is especially pronounced near the DM frequencies. In Fig. 8(d) the evolution of the transmittivity spectrum of the *x*-polarized EMW is given. We calculated the transmittivity profiles for the PC with the SC sublayer of fixed thickness d_s=10 nm, at temperatures *T=4.2 K*, *T=20 K*, *T=40 K*, *T=60 K*, *T=77 K*, and *T=90 K*. As follows from Fig. 8(d), the transmittivities of both LF and HF DMs drop down with increasing temperature, but the LF peak becomes visibly weaker than the HF one for temperatures above *T=77 K*. Meanwhile, the temperature practically does not affect the DM positions. At *T=4.2 K* the transmittivities of the LF and HF DMs are $T_{LF}^{(x)} \approx 0.42$ and $T_{HF}^{(x)} \approx 0.79$. At *T=20 K* they become $T_{LF}^{(x)} \approx 0.27$ and $T_{HF}^{(x)} \approx 0.65$, respectively. At *T=77 K* the LF peak drops to a vanishingly small value of $T_{LF}^{(x)} \approx 0.07$, and further it practically flattens at *T=90 K*, while the HF one still stays relatively large: $T_{HF}^{(x)} \approx 0.35$. One can see that the change of temperature produces only a slight effect on both PBG edges.

In Fig. 9(a) and 9(b) we present the transmitivity spectrum dependence on d_s for *y*-polarized DM, calculated for *T=4.2 K* (the top-view and profiles, respectively).

In contrast to the case of the *x*-polarized DMs given in Figs. 8(a) and 8(c), the transmittivity of the *y*-polarized DM practically does not depend on the SC defect sublayer thickness, when d_s is about several tens of nanometers.

The transmittivity spectra of the PCs are also sensitive to the thickness of the dielectric defect sublayer d_{1d}. In Figs. 9(a) and 9(b) the top-views of $T^{(x)}$ as function of the normalized thickness

http://dx.doi.org/10.5772/59971

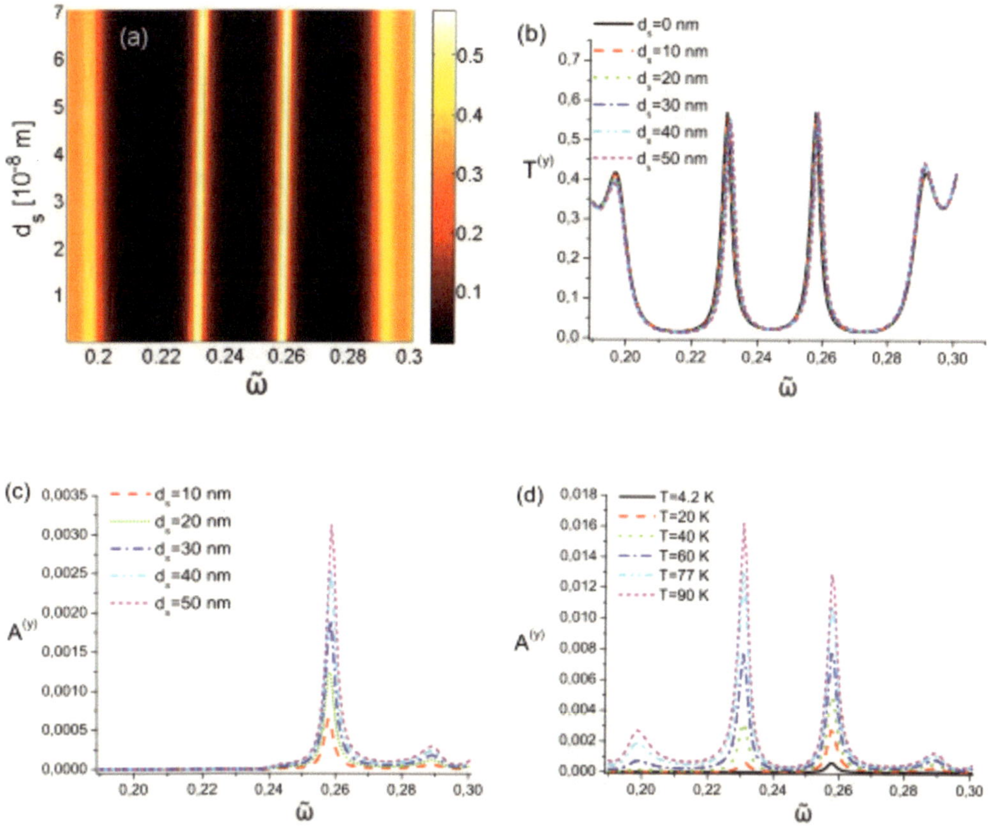

Figure 9. Two RH SC defects. Top view of the transmittivity $T^{(y)}$ vs normalized frequency $\widetilde{\omega}$ and defect SC sublayer thickness d_s for T=4.2 K (a); transmittivity $T^{(y)}$ (b) profiles for different d_s calculated for T=4.2 K. The results are obtained for the case of the dielectric defect layer with d_{1d}=d_1=2.1 μm.

of the dielectric sublayer d_{1d} /D and normalized frequency $\widetilde{\omega}$ are presented for the PCs with the SC defect sublayers thicknesses d_s=10 nm and d_s=20 nm. From Figs. 9(a) and 9(b) one can see that the thickness increase of both dielectric defect sublayers results in a shift of the DMs to the LF PBG edge. It should be mentioned, that the analogous increase of the SC defect sublayers d_s leads to a shift of the DMs in the opposite direction, i.e. to the HF PBG edge. This shift becomes noticeable in the scale of a few tens of percents of the period D.

The other peculiarity of the dependence of $T^{(x)}$ on d_{1d} is its periodicity. Comparing Figs. 10(a) and 10(b), one can see that the increase of d_s leads to a narrowing of the DMs and to their decaying (the HF DM decays stronger).

Above we investigated the PC with two identical complex (SC + dielectric) defects but with different thicknesses of SC sublayers. We can fix one of the SC sublayer thickness d_{sR} or d_{sL} (in the left or right combined defect, respectively) and vary the other SC sublayer thickness to obtain a possible changing of the PBG spectra. In Figs. 11(a) – 11(c) we present top views of

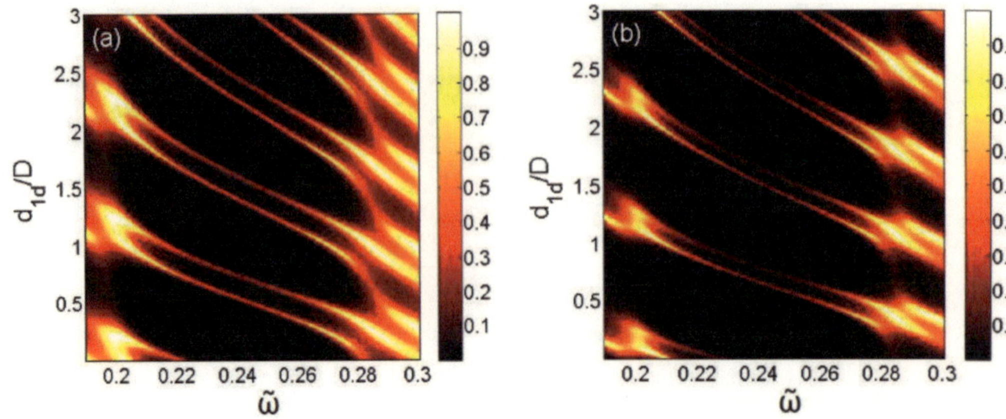

Figure 10. Two RH SC defects. Top view of the transmittivity $T^{(x)}$ vs normalized frequency $\tilde{\omega}$ and normalized thickness d_{1d}/D for d_s=*10 nm* (a); and d_s=*20 nm* (b). The calculations are performed for *T*=*4.2 K*.

the transmittivity of *x*-polarized EMWs $T^{(x)}$ as function of $\tilde{\omega}$ and d_{sR}, when the left-side SC defect thickness is fixed to d_{sL}=10 nm (a), d_{sL}=20 nm (b), and d_{sL}=30 nm (c).

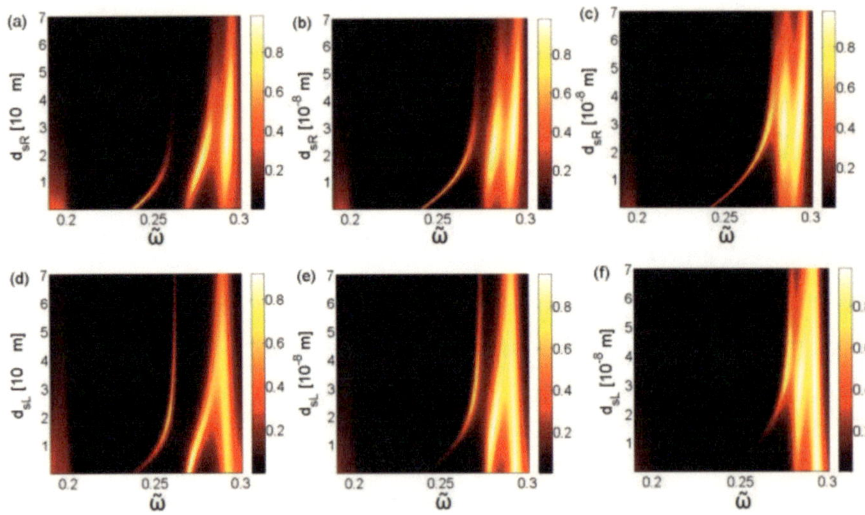

Figure 11. Two RH SC defects.Top view of transmittivity $T^{(x)}$ vs normalized frequency $\tilde{\omega}$ and thickness of the RH or LH SC defect sublayers (d_{sR} or d_{sL}, respectively). For the case of fixed LH SC defect layers with (a) d_{sL}=*10nm*, (b) d_{sL}=*20 nm, and* (c) d_{sL}=*30nm*; and for the case of fixed RH SC defect layers with (d) d_{sR}=*10nm*, (e) d_{sR}=*20nm*, and (f) d_{sR}=*30nm*. The calculations are performed for *T*=*4.2 K*, d_{1d}=d_1, and *N*=5, *M*=2.

The corresponding spectra for the opposite case (with fixed thickness of the right-side SC defect d_{sR}=10 nm, 20 nm, and 30 nm) are given in panels (d), (e), and (f), respectively. Comparing Figs. 8(a) and 11(a) for symmetrical and asymmetrical changes of the SC defect layers thickness, one can see that the behavior of the DMs changed. In contrast to the case of identical SC sublayers given in Figs. 8(a), for the case of fixed d_{sL}=10 nm the LF DM is more narrow and it decays fast with increasing of d_{sL}, not merging with the HF PBG edge as shown in Fig. 8(a). With the increase of the fixed value of d_{sL} to 20 nm (Fig. 11(b)) the HF DM comes closer to the HF PBG edge and merges with it for d_{sL}=30 nm for all values of d_{sR}. For d_{sL}=20 nm and d_{sL}=30 nm the LF DM grows with increasing of d_{sR} (Figs. 11(b) and 11(d)) and merges with the HF PBG edge for d_{sL}=30 nm at about d_{sR}=40 nm. Varying the left-side SC defect layer thickness one can obtain another dependence of the spectra on d_{sL} (Figs. 11(d) – 11(f)). For this situation the LF DM becomes visible for larger values of d_{sL}. The tendency of the shift of DM to higher frequencies with the increase of both d_{sL} and d_{sR} remains.

3.2. PC with two LH defects

In this subsection we present the numerical results for the PC with two LH defect layers depicted schematically in Fig. 7(b). Similar to the previous case, we study the EMW's incidence on the right-hand surface of the PC, and all the parameters of the PC are the same, as described in detail above in section 3.1, except for the replacing of the RH combined defects to LH ones.

In Figs. 12(a) – 12(d) we present the resulting transmittivities $T^{(x)}$ for the PC with two LH defects, analogous to that of Figs. 8(a) – 8(d).

In Figs. 12(a) and 12(b) one can see the top view of the transmittivities vs the normalized frequency and the defect SC sublayer thickness d_s for the temperatures T=4.2 K and T=77 K, respectively. The SC sublayer thickness d_s varies in the range (0 ÷ 70) nm, while the dielectric defect layer thickness is fixed to be d_{1d}=d_1=2.1 μm. In Fig. 12(c) we demonstrate the evolution of the transmittivity profiles $T^{(x)}$ with the increase of d_s.

Comparing the corresponding parts of Figs. 12 and 8, one can see that the behavior of the transmittivity $T^{(x)}$ has drastically changed. Though analogously to the case of the PC with two RH defects, both DMs have a tendency to shift to higher frequencies with the increase of d_s, now this shift is substantially smaller. The solid lines in Fig. 12(c) and 8(c) denote the transmittivity spectrum for d_s=0 nm and are equivalent. For d_s=10 nm the LF and HF peak positions become $\tilde{\omega}_{LF}^{peak}$ =0.23218 (which is equivalent to ω_{LF}^{peak} =87.47 THz or λ_{LF}^{peak} =21.53 μm) and $\tilde{\omega}_{HF}^{peak}$ =0.2524 (ω_{HF}^{peak} =97.66 THz or λ_{HF}^{peak} =19.28 μm), respectively. The frequency shifts of the LF and HF DMs are practically the same $\Delta\tilde{\omega}_{LF} \approx \Delta\tilde{\omega}_{HF}$ = 0.0016 (which is equivalent to $\Delta\tilde{\omega}_{LF}^{peak}$ =0,6 THz or $\Delta\lambda_{LF}^{peak}$ =0.15 μm) which is about 1.8% of the PBG width at d_s=0 nm. In contrast to the case of the 1D PC with two RH defects, the distance between the DMs practically does not change with the increase of the SC sublayer thickness remaining about $\Delta\tilde{\omega}^{peak}$ ≈0.0271 ($\Delta\omega^{peak}$ ≈10.21 THz or λ^{peak} ≈2.25 μm) for d_s varying within the range of (0 ÷ 70) nm. Moreover, in this geometry the DMs do not merge with the HF PBG edge when d_s varies within some tens of nanometers, as was obtained in the case with double RH defects. In addition, the LF PBG edge splits when d_s exceeds the value of about 10 nm and then one branch significantly approaches the LF DM with increasing of d_s and merges with it at d_s≈40 nm. The intensity of the LF DM increases with the increase of d_s, from T=0.6 for the case of purely dielectric defect layers (d_s=0

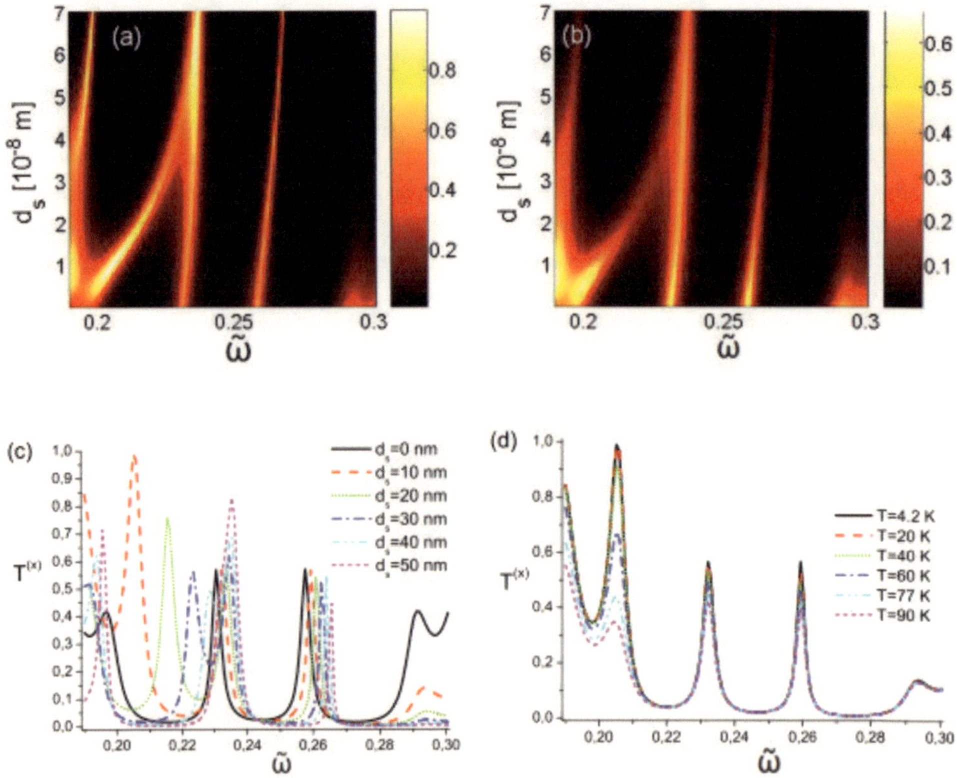

Figure 12. Two LH SC defects. The same as in Fig. 8, except for the case of LH SC defects.

nm) to $T=0.85$ for d_s=50 nm. The HF PBG edge decays fast with the increase of d_s, in contrast to the case of RH defects, where the maximum values of $T^{(x)}$ are achieved at the HF PBG edge.

In Fig. 12(d) one can see the evolution of the transmittivity spectra of the first PBG with changing temperature. The results are given for the PC with the SC defect sublayers and the dielectric defect ones d_s=10 nm and d_{1d}=2.1 μm, respectively. As follows from Fig. 12(d), the maximum influence of the temperature on the $T^{(x)}$ spectrum takes place at the LF PBG edge, where the transmittivity drops from $T^{(x)}_{LF,\,PBG}$ =0.99 at $T=4.2$ K to $T^{(x)}_{LF,\,PBG}$ =0.34 at $T=90$ K. The behavior of the DM's peaks with temperature increase is similar, but much less pronounced: the transmittivity slightly diminishes from $T^{(x)}_{LF}$ ≈0.57 at $T=4.2$ K to $T^{(x)}_{LF}$ =0.42 and $T^{(x)}_{HF}$ =0.40. The HF PBG edge practically does not experience any affect of the temperature variation within the interval from 4.2 K to 90 K.

As for the behavior of the transmittivity of the y-polarized light $T^{(y)}$, it practically does not depend on the type of defects (RH or LH) in the PC, and its variation with increasing temperature and d_s remains similar as depicted in Fig. 9(a) for the PC with two RH defects.

Examining the transmittivity spectra dependence on the normalized thickness of the dielectric sublayer d_{1d}/D, analogously to the case investigated above for the 1D PC with two RH defects, we obtained the periodically repeating structures in the top views of $T^{(x)}$ vs $\tilde{\omega}$ and d_{1d} (Figs. 13(a) and 13(b)). As before, the spectra variations with d_{1d} are calculated for the SC defect sublayer fixed to be d_s=10 nm (Fig. 13(a)) and d_s=20 nm (Fig. 13(b)). The thickness increase of both dielectric defect sublayers results in a shift of the DMs to the LF PBG edge, as it was obtained above for a 1D PC with two RH defects. But for a 1D PC with two LH defects, one can see the distortion of the DMs on the LF PBG edge with the increase of d_s to 20 nm. A further enlargement of d_s leads to the DM's lines breaking.

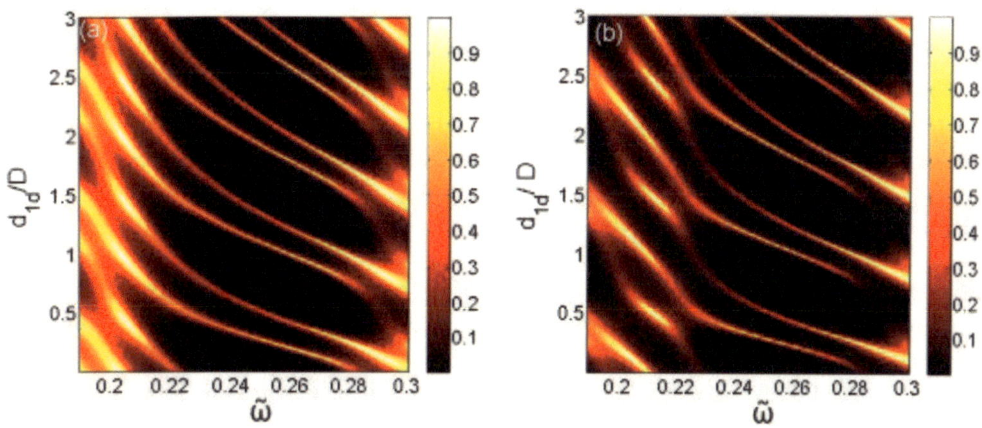

Figure 13. The same as for Fig. 11, except for the case of two LH SC defects.

In Figs. 14(a) – 14(c) we show the top views of transmittivity $T^{(x)}$ versus $\tilde{\omega}$ (within the 1st PBG) and SC defect layer thicknesses d_{sR} (left SC defect sublayer). Varying d_{sR} from 0 to 70 nm we keep the thickness of the left-sided SC defect sublayer fixed to be (a) d_{sL}=10 nm, (b) d_{sL}=20 nm, and (c) d_{sL}=30 nm. The calculations of $T^{(x)}$, presented in Figs. 14(d) – 14(f), are performed for T=4.2 K and dielectric defect sublayers of thickness d_{1d}=d_1=2.1 μm for a PC structure with the period numbers N=5 and M=2. Comparing Figs. 14(a) – 14(c) with Fig. 12(a), we see that in the case of asymmetrical SC defects the LF PBG edge does not merge with the LF DM when we change the right-sided SC defect sublayer thickness d_{sR}. Varying d_{sL} while d_{sR} is fixed we obtain a mergence of the decreasing LF DM with the LF PBG edge for d_{sR}=30 nm, similar to that in Fig. 12(a) for $d_{sL} \approx 60$ nm (Fig. 14(f)).

Considering asymmetrical changes of the SC defect sublayer thicknesses, as it was done analogously for the case of two RH defects in section 3.1, for the case of a 1D PC with two LH we obtain different spectra when changing one of the SC defect thicknesses while the other one is fixed.

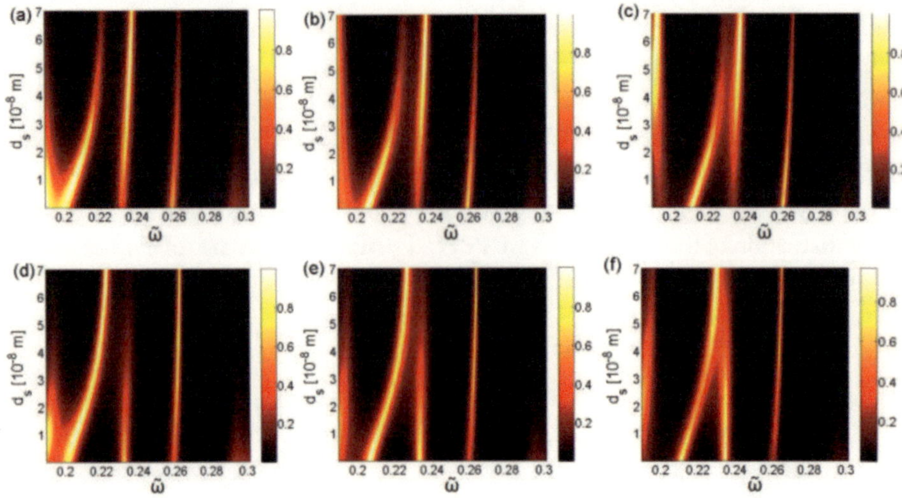

Figure 14. The same as for FIG. 11, except for LH – LH geometry.

3.3. PC with RH — LH and LH — RH defects

In this subsection we present the numerical results for the PCs with two defect layers depicted schematically in Fig. 7(c) and (d). In Figs. 15(a) and 15(b) we present the top view of the transmittivities $T^{(x)}$ for the x-polarized EMWs vs the normalized frequency $\tilde{\omega}$ and the defect SC sublayer thickness d_s for RH – LH and LH – RH geometries, respectively.

The SC sublayer thickness d_s varies within the range (0 ÷ 70) nm, while the dielectric defect layer thickness is fixed to be d_{1d}=d_1=2.1 μm. The color in Figs. 2(a) and 2(b) denotes the value of the transmittivity, as shown on the panels. The results are obtained for temperature T=4.2 K. The presence of two combined defect layers in the PC leads to appearance of two narrow DMs inside the PBG (see Figs. 15(a), 15(b)) which further are referred to as low-frequency (LF) and high-frequency (HF) DMs.

As one can see from Figs. 15(a) and 15(b), for both geometries (RH – LH and LH – RH), the increase of d_s from 0 to 70 nm leads to a shift of both LH and HF DMs to higher frequencies, but the shifts of HF DMs are larger than of the LF ones.

In the case of RH – LH geometry, first the LF defect mode slightly deviates to higher frequencies with increasing d_s and further, after d_s≈20 nm, the LF DM gets thin and its position does not change with further increase of d_s, while the HF DM peak merges with the HF PBG edge and the transmittivity values of both HF PBG edge and HF DM decrease. For all values of d_s from the considered interval (except d_s=0 nm) the LF defect mode is more pronounced then the HF one.

For the case of LH – RH geometry, the behavior of the defect modes is opposite: the LF defect mode is less pronounced than the HF one. For d_s=10 nm the LF DM is about 8 times smaller in

http://dx.doi.org/10.5772/59971

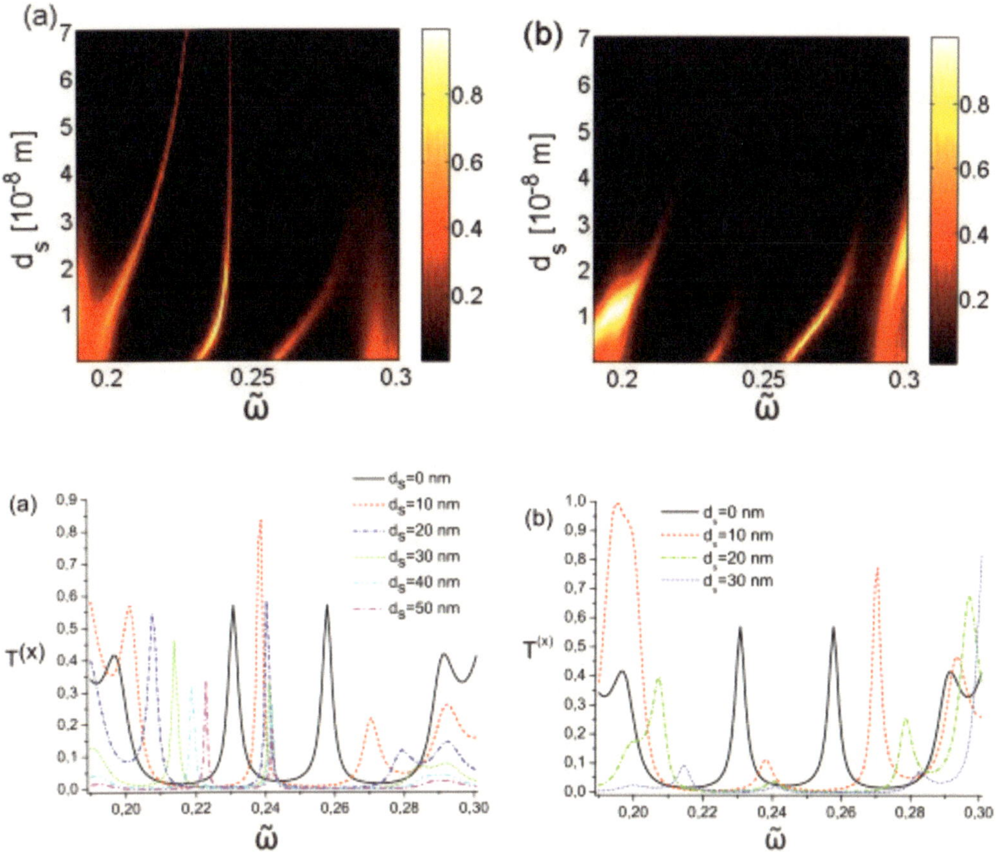

Figure 15. Top view of the transmittivity $T^{(x)}$ vs normalized frequency $\widetilde{\omega}$ and the defect SC sublayer thickness d_s for the cases of RH – LH (a) and LH - RH defect layers (b). The transmittivity spectra $T^{(x)}$ for different values of the SC sublayer thicknesses for the RH – LH (a) and LH - RH geometries (b). The results are obtained for the case of the dielectric defect layer d_{1d}=2.1 μm and temperature T=4.2 K.

magnitude then the HF DM, and for d_s>20 nm the LF DM is practically suppressed, while the HF DM still exists up to d_s=30 nm.

In Figs. 15(c) we give the transmittivity $T^{(x)}$spectra for d_s=0 (solid lines), d_s=10 nm (dashed lines), d_s=20 nm (dash-dotted lines), d_s=30 nm (dotted lines), d_s=40 nm (dash-dot-doted lines), and d_s=50 nm (long dash-dotted lines) for the case of RH – LH geometry. In the case of LH – RH geometry, for d_s increasing 30 nm the both DMs are suppressed and in Fig. 15(d) we show the $T^{(x)}$ spectra only for d_s=0 (solid lines), d_s=10 nm (dashed lines), d_s=20 nm (dash-dotted lines), and d_s=30 nm (dotted lines). Obviously, in the case of pure dielectric defects (d_s=0) the spectra for RH – LH and LH – RH geometries are identical (the solid lines in Figs. 15(c) and 15(d)). In this case two DM peaks of equal values $T_{LF}^{(x)} = T_{HF}^{(x)} \approx 0.57$ are localized symmetrically inside the PBG at $\widetilde{\omega}_{LF}^{peak}$ =0.2307 and $\widetilde{\omega}_{HF}^{peak}$ =0.2577 which correspond to ω_{LF}^{peak} =86.91 THz and ω_{HF}^{peak} =97.08

THz, respectively. For d_s=0 the positions of the PBG edges are $\widetilde{\omega}_{LF}^{PBG}$ =0.1967 (ω_{LF}^{PBG} =74.10 THz) and $\widetilde{\omega}_{HF}^{PBG}$ =0.2915 (ω_{HF}^{PBG} =109.82 THz), so the PBG width is $\Delta\widetilde{\omega}_{PBG}$ =0.0948 ($\Delta\omega_{PBG}$ =35.71 THz), and the transmittivity values at both PBG edges are equal: $T_{LF}^{(x)PBG} = T_{HF}^{(x)PBG}$ ≈0.41.

The introduction of the SC sublayers into a PC with two dielectric defect layers changes the symmetry of the positions and values of DMs, as well as of the PBG edges. For example, the SC sublayers of thicknesses d_s=10 nm drastically change the transmittivity spectra. As one can see from Figs. 15(c) and 15(d), both DM peaks shifts to higher frequencies. For the RH – LH geometry at d_s=10 nm the DM peaks are located at $\widetilde{\omega}_{LF}^{peak}$ =0.2385 (ω_{LF}^{peak} =89.85 THz) and $\widetilde{\omega}_{HF}^{peak}$ =0.2705 (ω_{HF}^{peak} =101.90 THz). So, the LF DM shift from its position at d_s=0 is $\Delta\widetilde{\omega}_{LF}^{peak}$ ≈0.0078 ($\Delta\omega_{LF}^{peak}$ ≈2.94 THz), while the HF DM shit is $\Delta\widetilde{\omega}_{HF}^{peak}$ ≈0.0128 ($\Delta\omega_{HF}^{peak}$ ≈0.48 THz). The distance between the DM peaks becomes $\Delta\omega^{peak}$ ≈0.0317 (approximately 11.94 THz), which is by 17% larger than for the case of pure dielectric defect layers.

For the RH – LH geometry increasing d_s leads to a shrinking of the PBG width: the LH PBG edge shifts to higher frequencies, while the HF PBG edge remains at its position. So, the PBG width for d_s to 10 nm is $\Delta\widetilde{\omega}_{PBG}$ ≈0.08063, which is about 30.38 THz which is 15% of the PBG width at d_s=0. One can be mentioned that for both RH – LH and LH – RH geometries the PBG spectra variation with increasing d_s is quite different from the cases described in the previous subsection, where we considered the asymmetrical configurations RH – RH and LH – LH.

In Figs. 16(a) and 16(b) we show the transmittivity $T^{(x)}$ evolution with changing temperature for the PCs of RH – LH and LH – RH geometries, respectively. The spectra are calculated for the defect sublayer thicknesses d_s=10 nm and d_{1d}=d_1=2.1 μm.

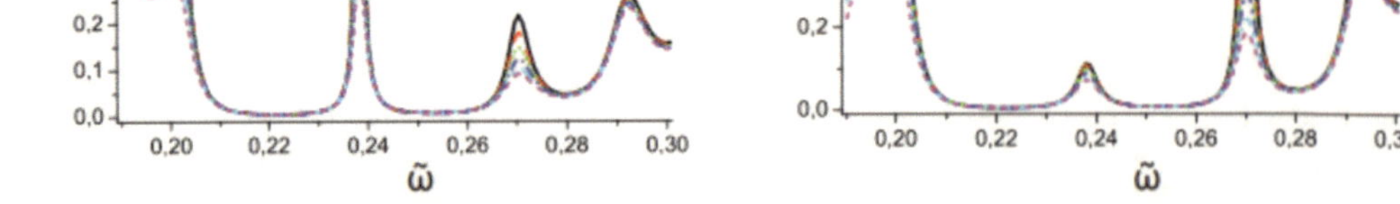

Figure 16. The temperature evolution of transmittivity spectra for the PC with RH – LH (a) and (b) LH – RH complex SC defects. The results are obtained for the defect layer thicknesses d_s=10 nm and d_{1d}=d_1=2.1 μm.

For the case of RH – LH geometry (Fig. 16(a)) the temperature increase leads to the abrupt fall of the transmittivity value $T^{(x)}$ at the LF defect mode $\widetilde{\omega}_{LF}$ =0.23855 from $T_{LF}^{(x)}$ ≈0.83 at *T=4.2 K* to $T_{LF}^{(x)}$ ≈0.31 at *T=90 K*. As well, the HF DM peak diminishes with increasing temperature from

$T_{HF}^{(x)}$ ≈0.31 at *T*=4.2 *K*, to $T_{HF}^{(x)}$ ≈0.09 at *T*=90 *K*. In Fig. 16(a) one can see an essential lowering of the LF PBG edge in contrast to the HF PBG edge, which practically does not change with temperature. In the case of LH – RH geometry (Fig. 16(b)) the temperature effects substantially on the HF defect mode: its magnitude drops more than four times from $T_{HF}^{(x)}$ ≈0.775 at *T*=4.2 *K* to $T_{HF}^{(x)}$ ≈0.187 at *T*=90 *K*. The changing of the LF DM is not essential: at *T*=4.2 *K* the LF DM peak is quite small ($T_{LF}^{(x)}$ ≈0.118) and at *T*=90 *K*, it diminishes for 39% of this value and becomes $T_{LF}^{(x)}$ ≈0.072. At liquid helium temperature the transmittivity $T^{(x)}$ at the LF PBG edge is very high $T_{LF}^{(x)PBG}$ ≈0.996 and with the increase of temperature to T_C it becomes more than two times smaller. The lowering of the HF PBG edge with temperature is not so pronounced ($T_{HF}^{(x)PBG}$ for *T*=90 *K* is 19% lower than for *T*=4.2 *K*).

Our numerical results demonstrate that the PBG spectra of the *y*-polarized waves practically do not change when d_s is about several tens of nanometers, in contrast to the case of the *x*-polarized EMWs. The analogous result was obtained in our previous calculations for the PCs with two RH and two LH combined SC defects in the previous subsection. In Fig. 17 we show the top-view of the transmittivity $T^{(y)}$ of *y*-polarized EMW as function of normalized frequency $\tilde{\omega}$ and the SC defect sublayer thickness d_s for RH – LH geometry. The corresponding dependence for LH – RH geometry looks similarly. As one can see form Fig. 17, the positions of both DM peaks and both the positions and values of the PBG edges stay practically without changes with increase of d_s. The variation of the DM transmittivity with d_s is quite small for the *y*-polarized EMWs in comparison with the *x*-polarized ones.

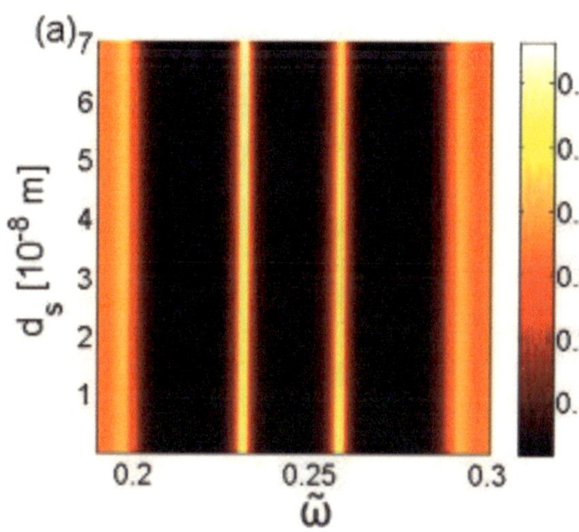

Figure 17. Top view of the transmittivities $T^{(y)}$ as function of $\tilde{\omega}$ and d_s (a). The results are obtained for the dielectric defect layer with d_{1d}=2.1 μm for *T*=4.2 *K*.

In contrast to the x-polarized EMWs, the spectra of the y-polarized EMWs practically do not exhibit noticeable modification with temperature for both (RH – LH and LH – RH) geometries. According to our estimations, for RH – LH geometry, the lowering of the transmittivities at frequencies of the LF DM and HF DM (with temperature increasing from T=4.2 K to T=77 K) is about 2% and 1.7% of their corresponding values for T=4.2 K, respectively. For the LH – RH geometry the lowering of DM peaks are 1.9 for the LF DM and 1.72% for the HF DM.

Above we investigated the PC with two combined defects (of RH – LH or LH – RH geometries) and supposed the equal thicknesses of the SC sublayers. Also we can fix one of the SC sublayer thicknesses d_{sR} or d_{sL} (in the left or right combined defect layer, respectively) and vary the other SC sublayer thickness, one can obtain a changing of the PBG spectra. Further we refer to this case, as an asymmetrical change of the SC defect sublayers. We start the investigation of the asymmetrical change of the SC defect sublayer thickness for the PC of RH – LH geometry. In Figs. 18(a) – 18(c) we present the top views of transmittivity $T^{(x)}$ of x-polarized EMWs as function of $\tilde{\omega}$ and d_{sR}, with the fixed left-sided SC defect thicknesses d_{sL}=10 nm (a), d_{sL}=20 nm (b), and d_{sL}=30 nm (c). Comparing Figs. 15(a) and 19(a) – (c) for symmetrical and asymmetrical changes of the SC defect layers thickness, one can see, that the behavior of the DMs is changed. In contrast to the case of the equal thickness SC sublayers given in Fig. 15(a), for the case of fixed d_{sL}=10 nm both LF and HF DMs are practically do not change their positions with the increase of d_{sR}. Moreover, the HF DM is more broadened and it merges with the HF PBG edge completely for all d_{sR} for d_{sL}=30 nm (see Fig. 18(c)), so in this case we have only one DM inside the PBG. Comparing the transmittivity of the LF DM for the same values of d_{sR} at the Figs. 18(a) – (c), one can see that $T^{(x)}$ diminishes with the increase of d_{sL}.

Varying the left-sided SC defect layer thickness d_{sL} with fixed d_{sR}, one can obtain another behavior of the spectra with fixed thickness d_{sR}=10, 20 and 30 nm) are given in Figs. 18(d), 18(e), and 18(f), respectively. For this case a shift of both DMs to higher frequencies remains for d_{sR}=10, 20 and 30 nm, similar to the case of symmetric change of d_s which is shown in Fig. 15(a). But with the increase of the fixed value of d_{sR} to 20 nm the LH PBG edge peak sharpens and for d_{sL}=30 nm it splits of PBG transforming to the new pronounced DM, while the transmittivity of the HF DM as well as of the FH PBG edge, decreases with increasing of d_{sR}. In Fig. 18(f) both the new DM and the LF DM slightly deviate to higher frequencies with the increasing d_{sL}.

Analogously, in the case of the PC of the LH – RH geometry, the asymmetric changing of the SC defect sublayer thickness modifies the dependence of the transmittivity specter given in Fig. 15(b) for the symmetric changing of d_s. As one can see from Figs. 19(a) – 19(c), for the fixed values of d_{sL}, the LF PBG edge does not shift to higher frequencies with the increasing d_{sR}, as in Fig. 15(b). The LF DM decreases fast with the increase of d_{sR} and d_{sL}.

Varying d_{sL} for the fixed values of d_{sR}, one cannot shift the LF DM to higher frequencies, as shown in Figs. 19(d) – (e), but the transmittivity of the LF DM decreases sharply with the increase of d_{sL} and it is practically suppressed for d_{sR}=30 nm. For d_{sR}=30 nm the HF DM merges with HF PBG edge for small values of d_{sL}.

http://dx.doi.org/10.5772/59971

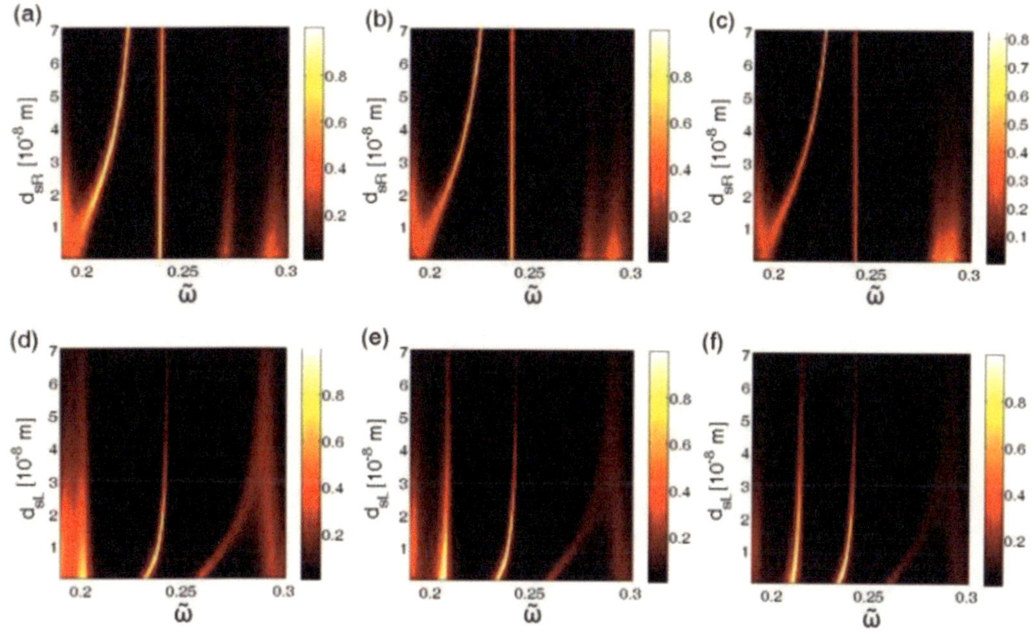

Figure 18. RH – LH geometry. Top view of the transmittivities $T^{(x)}$ vs $\widetilde{\omega}$ and d_s for the case of fixed left-sided SC defect thickness: (a) d_{sL}=10 nm, (b) d_{sL}=20 nm, and (c) d_{sL}=30 nm ; and for the case of fixed right-sided SC defect thickness:(d) d_{sR}=10 nm, (e) d_{sR}=20 nm, and (f) d_{sR}=30 nm. The calculations are performed for T=4.2 K, d_{1d}=d_1.

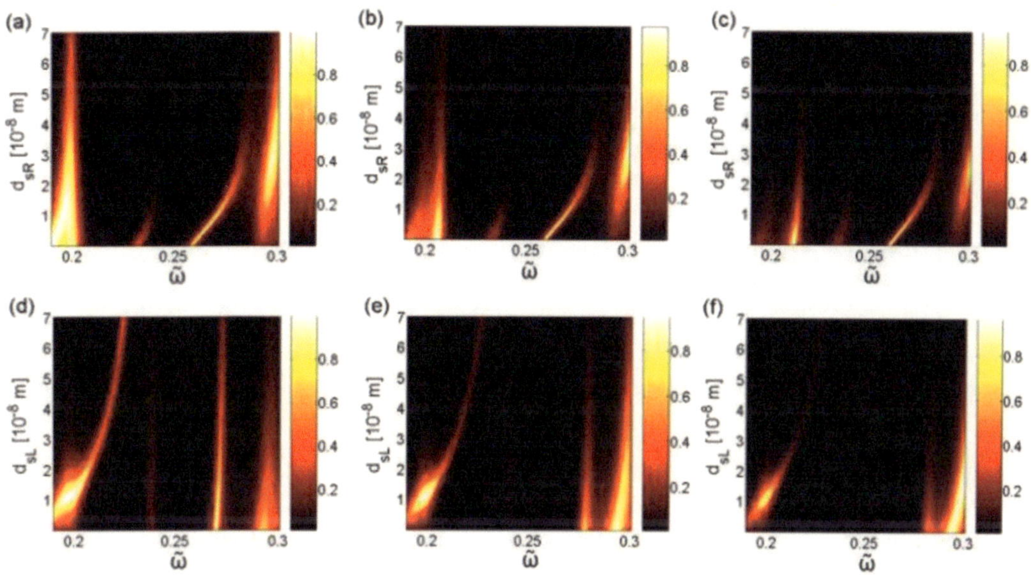

Figure 19. The same as for Fig. 9, except for LH – RH geometry.

4. Conclusions

In conclusion, we have investigated the behavior of DMs in a 1D dielectric PC with two complex bilayer defects composed of SC and dielectric constituents. We have considered the case of a fixed distance between the two defects embedded into the PC asymmetrically, i.e., when both SC sublayers are located on the right-hand side (RH – RH geometry) or on the left-hand side (LH – LH geometry) with respect to the dielectric parts of the complex defects. The normal incidence of linearly polarized EMWs (with electric field vector **E** perpendicular to the axis of the PC's growth direction) on the PC is investigated.

The positions of the transmittivity peaks at the DM frequencies inside the first PBG are studied both analytically and numerically, for different temperatures and thicknesses of SC and dielectric sublayers. We have shown that the increase of temperature from liquid helium temperature (T=$4.2\ K$) to the critical temperature of the SC sublayer (T_C=$90\ K$ for $YBa_2Cu_3O_7$) leads to significant changes of the DM's transmittivity peaks in the PBG spectra of x-polarized EMWs, up to a practically complete suppression of one of them (the low-frequency DM for RH – RH geometry), while the spectra of y-polarized EMWs are practically unchanged. The pronounced contrast in behavior of x- and y-polarized modes is based on the in-plane anisotropy of the dielectric tensor components of the SC sublayer. The positions of PBG edges and DM's remain invariable with temperature for both EMW's polarizations. We have demonstrated a high sensitivity of the PBG spectra of x-polarized EMWs to variations of the SC defect sublayers thicknesses in both RH – RH and LH – LH geometries. The increase of the SC defect sublayer thickness leads to a shift of both DM peaks towards the high-frequency PBG range up to their mergence with it, as well as to a decrease of the distance between these peaks (for RH – RH geometry), or a substantial shift of the position of the low-frequency PBG edge to higher frequencies (LH – LH geometry).

We have also considered the case of a fixed distance between the two defects symmetrically embedded into the PC, i.e., when one of the SC sublayers is located on the right-hand side and the another one on the left-hand side with respect to the dielectric parts of the complex defects (the cases of RH – LH and LH – RH geometries). The increase of the SC defect sublayer thickness leads to a shift of both DM peaks towards the high-frequency PBG range up to mergence of the one of them with it (in the RH – LH geometry), as well as to an increase of the distance between these peaks, or a substantial shift of the position of the low-frequency PBG edge to higher frequencies (in both RH – LH geometries). The variation of the dielectric defect sublayer's thicknesses allows shifting the DM peaks in spectra of both x- and y-polarized EMWs towards the low-frequency PBG edge for each geometry (RH – RH, LH – LH, RH – LH, and LH – RH).

Changing the SC defect sublayer thicknesses asymmetrically (when the right or the left SC defect sublayer's thickness is fixed while the another one is varied), one can modify the PBG spectra of x-polarized EMWs, changing the number of the defect modes in the PBG with chosen thicknesses of the SC defect sublayers.

As follows from our numerical results, the dielectric PCs with SC defects can be used as constituents of polarization-selective narrow-band filters for THz radiation [31, 32]. The optimal parameters of these filters can be obtained choosing the proper SC and dielectric defect sublayer thicknesses. The high sensitivity to temperature of the PCs with SC defects opens possibilities to use such a structures as the basis of temperature tunable electromagnetic filters in the THz regime [33].

Author details

N. N. Dadoenkova[1,2], A. E. Zabolotin[1], Yu. S. Dadoenkova[1,2], I. L. Lyubchanskii[1*], Y.P. Lee[3] and Th. Rasing[4]

*Address all correspondence to: igorl@fti.dn.ua

1 Donetsk Physical & Technical Institute of the National Academy of Sciences of Ukraine, Donetsk, Ukraine

2 Ulyanovsk State University, Ulyanovsk, Russian Federation

3 q-Psi and Department of Physics, Hanyang University, Seoul, Republic of Korea

4 Radboud University Nijmegen, Institute for Molecules and Materials, Nijmegen, The Netherlands

References

[1] Yablonovich E. Physical Review Letters 1987; 58: 2059.

[2] John S. Physical Review Letters 1987; 58: 2486.

[3] Jannopoulos JD, Johnson SG, Winn JN, and Meade RD. Photonic Crystals: Molding the Flow of Light, 2nd ed. Princeton, New York: Princeton University Press; 2008.

[4] Sakoda K. Optical Properties of Photonic Crystals. Berlin: Springer; 2005.

[5] Vittorio M.N. Passaro., editor. Advances in Photonic Crystals (Intechopen); 2013.

[6] Shabanov VF, Vetrov SY, and Shabanov AV. Optics of real photonic crystals. Mesomorphic Defects and Inhomogeneities. Novosibirsk: Russian Academy of Sciences; 2005 (in Russian).

[7] Lyubchanskii I L, Dadoenkova N N, Lyubchanskii M I, Shapovalov E A, and Rasing Th. Journal of Physics D: Applied Physics 2003; 36: R277.

[8] Inoue M, Fujikawa R, Baryshev A, Khanikaev A, Lim P. B, Ushida H, Aktsipetrov O, Fedyanin A, Murzina T, and Granovsky A. Journal of Physics D: Applied Physics 2006; 39: R151.

[9] Scott JF. Ferroelectrics 2003; 293: 33.

[10] Fu Y, Zhang J, Hu X, and Gong Q. Journal of Optics 2010; 12: 075202.

[11] Ferrini R. Liquid Crystals into Planar Photonic Crystals. In: Tkachenko GV. (ed.) New Development in Liquid Crystals. Rijeka: InTex; 2009. p21-48.

[12] Da H-X and Li ZY. Manipulating Nematic Liquid Crystals-based Magnetophotonic Crystals. In: Tkachenko GV. (ed.) New Development in Liquid Crystals. Rijeka: InTex; 2009. p50-70.

[13] Savel'ev S, Rakhmanov AL, and Nori F. Physical Review Letters 2005; 94: 157004.

[14] Savel'ev S, Rakhmanov AL, and Nori F. Physical Review B 2006; 74: 184512.

[15] Savel'ev S, Yampol'skii VA, Rakhmanov AL, and Nori F. Repts. Progr. Phys. 2010; 73: 026501.

[16] Anlage SM. Journal of Optics 2011; 13: 024001.

[17] Lyubchanskii IL, Dadoenkova NN, Zabolotin AE, Lee YP, and Rasing Th. Journal of Optics A: Pure and Applied Optics 2009; 11: 114014.

[18] Dadoenkova NN, Zabolotin AE, Lyubchanskii IL, Lee YP, and Rasing Th. Journal of Applied Physics 2010; 108: 093117.

[19] Rauh H and Genenko YA. Journal of Physics of Condensed Matter 2008; 20: 145203.

[20] Barvestani J. Optics Communications 2011; 284: 231.

[21] Bacerra G, Moncada-Villa OE, Granada JC. Journal of Superconductivity and Novel Magnetism 2012; 25: 2163.

[22] Wu J, Gao J. Holography, Proceedings of SPIE 2012; 8556: 85561Q.

[23] Hu C-A, Liu J-W, Wu C-J, Yang T-J, Yang S-L. Solid State Communications 2013; 157: 54.

[24] Steel MJ, Levy M, and Osgood RM. Journal of Lightwave Technology 2000; 18: 1297.

[25] Steel MJ, Levy M, and Osgood RM. Journal of Lightwave Technology 2001; 19: 1964.

[26] Lyubchanskii IL, Dadoenkova NN, Lyubchanskii MI, Sapovalov EA, Zabolotin AE, Lee YP, Rasing Th. Journal of Applied Physics 2006; 100: 096110.

[27] Lyubchanskii IL, Dadoenkova NN, Zabolotin AE, Lee YP, Rasing Th. Journal of Applied Physics 2008; 103: 07B321.

[28] E. D. Palik. Handbook of Optical Constants of Solids. New York: Academic Press; 1991.

[29] Berreman DW. Journal of the Optical Society of America 1972; 62: 502.

[30] Borisov SB, Dadoenkova NN, and Lyubchanskii IL. Optics and Spectroscopy 1993; 74: 670.

[31] Gennaro ED, Zannini C, Savo S, Andreone A, Masullo MR, Castaldi G, Gallina I and Galdi V. New Journal of Physics 2009; 11: 113022.

[32] Raum M, Li J-S, and Padilla WJ. Journal of Infrared, Millimeter, and Terahertz Waves 2013; 34: 1.

[33] Liu J-W, Chang T-W, Wu C-J. Filtering Properties of Photonic Crystal Dual-Channel Tunable Filter Containing Superconducting Defects. Journal of Superconductivity and Novel Magnetism 2014; 27: 67-72.

Chapter 2

Supercontinuum Generation With Photonic Crystal Fibers and Its Application in Nano-imaging

Shuanglong Liu, Wei Liu and Hanben Niu

Additional information is available at the end of the chapter

http://dx.doi.org/10.5772/59987

1. Introduction

Supercontinuum (SC) generation describes the phenomenon that the output spectrum becomes much wider than the input spectrum when an ultrashort optical pulse passes through a nonlinear medium, such as bulk fused silica, photonic crystal fiber and so on. Since this phenomenon was first discovered by Alfano and Shapiroin in 1970s by passing picosecond pulses through a bulk BK7 glass [1], it has been explored in a wide variety of nonlinear media, including solids, gas, inorganic liquids and various waveguides.

In 1974, an ingenious idea of introducing microstructures into traditional step index optical fibers to modify the guidance properties by Kaiser and Altle [2] established the theoretical foundation for photonic crystal fiber (PCF). And the exploration for fabricating such fibers were never stopped until the advent of an all-silica PCF in Southampton University by Russell and co-workers in 1996 [3]. His birth drew much more interest from various scientific researchers and initiated a revolution in SC generation.

The PCF used for SC generation often has a solid core surrounded by regularly arranged microscopic air holes running along the fiber length [4], resulting in a surrounding region with lower refractive index compared to that of the central district. Therefore the light will be trapped and guided in the fiber core through total internal reflection. This modified surrounding shows good flexibility to engineer the waveguide properties. It is reported that the zero-dispersion wavelength (ZDW) in fused silica can shift to a shorter wavelength than the intrinsic ZDW about 1300 nm by suitable design of the air hole diameter and air filling fraction [5]. Then Ranka *et al.* first reported that the PCF can exhibit anomalous dispersion at visible region [6]. And they obtained an ultrabroadband continuum generation from 400 nm to 1600 nm by injecting a femtosecond pulse into a 75 cm section PCF. In 2000, Wadsworth *et al.* observed soliton effects by pumping a short PCF near the ZDW in the anomalous regime [7]. Accompanied with other nonlinear effects, the fundamental and second soliton were identified. The phenomenon of soliton self-frequency shift [8] in a tapered air–silica microstructure fiber was reported by Liu *et al.* from 1.3 to 1.65 μm. Numerical modeling of pulse evolution in PCFs was first proposed by Husakou and Herrmann [9] based on reduced Maxwell equations. And the important roles of soliton fission, four-wave mixing in spectral broadening were first discussed in detail. Due to the lack of Raman scattering, much fine structure was lost in the solution. An extended nonlinear Schrödinger equation (NLSE) was first applied to the modeling of femtosecond pulse propagation in PCF [10] with higher-order dispersion, self-steepening and Raman effects included. The numerical simulations based on extended NLSE is accurately consistent with experimental results. Based on the previous works, Genty *et al.* quantitatively studied the mechanisms of SC generation when the PCF was pumped in its normal and anomalous regimes [11]. Soon afterwards, continuum generation from 1065 to 1375 nm was demonstrated by pumping a honey fiber with continuous wave [12]. These numerical and experimental studies led to a detailed description of the SC generation in PCFs from the ultraviolet to the infrared in both femtosecond, picosecond regimes and continuous wave.

SC generation in PCFs involves several nonlinear effects such as self-phase modulation, (SPM) cross-phase modulation (XPM), self-steepening and optical shock formation, Raman scattering, solition fission, four-wave mixing and so on. Greatest spectral broadening was reported when a PCF was pumped in the anomalous dispersion regime near the ZDW [13]. The broadening mechanism in this case is dominated by soliton dynamics, especially breakup of higher-order solitons into fundamental solitons and non-solitonic radiation, though the SPM effect in initial propagation. The ejected solitons will shifts toward longer wavelength under intrapulse Raman scattering, known as soliton self-frequency shift (SSFS). Spectral broadening in normal dispersion region is mainly due to SPM effect and optical wave breaking. With soliton dynamics suppressed in this region, the generated SC possesses a good temporal structure in spite of a smaller spectral broadening. Under specific pump pulses, continua with broad spectral bandwidth, good spatial coherence, uniform intensity profile and simple temporal structure can be generated in certain PCFs. These continua show significant applications in many fields such as optical frequency metrology, optical communication and cellular biology.

Recent years have seen the fast development of cellular biology based on optical microscope and coherent lasers. And many microscopies were developed to observe the fine structures inside biologic cells, such as scanning electron microscope [14] (SEM), scanning tunneling microscope [15] (STM), and atomic-force microscope [16] (AFM). Though they all have very high spatial resolutions, vacuum circumstance needed in SEM makes it not suit for live-cell imaging while STM and AFM are restricted to surface mapping due to the cantilever tip [17]. With the development of several novel microscopic methods, such as photo-activated localization microscopy [18] (PALM), stochastic optical reconstruction microscopy [19] (STORM), stimulated emission depletion [20] (STED) microscopy, fluorescence microscopies obtain a spatial resolution about 20 nm and are widely used in cellular biology. But photo-toxicity, photo-bleaching and influence of invasive marker on cells cannot be ignored in vivo imaging, let alone some biomolecule that are difficult to be labeled [21]. Therefore it is urgent to develop a microscopy with nano-scaled resolution to study the process of metabolism in live cells and its response to invasive substances.

As a new kind of microscopies based on molecular vibrations, coherent anti-Stokes Raman scattering (CARS) microscopy has exhibited their prospective applications in live-cell imaging for its distinct characteristics, including high sensitivity and spatial resolution, label-free chemical specificity, three-dimension sectioning capability [22-24]. In traditional CARS microscopy, the pump and Stokes pulses can resonant only single or few molecular bonds due to the limitation of the spectral bandwidth. Obviously it is not adequate to acquire accurate recognition and mapping of an unknown biomolecule or that with complex components. Although this can be achieved by sequentially tuning the frequency difference of them or synchronizing two ultrashort lasers, it is time-consuming and costs a lot. When a SC pulse generated in a high nonlinear PCF serves as both the pump the Stokes, most distinct vibration modes, perhaps full modes even, can be probed simultaneously as long as different frequency components in the generated SC propagate at about a same speed. And a method named additional probe-beam-induced phonon depletion [25] (APIPD) is proposed to improve the spatial resolution on CARS microscopy. In APIPD method, an additional doughnut beam with different frequency from that of the probe beam is introduced to deplete the phonons on the periphery of point spread function (PSF). The subsequent Gaussian probe beam reacts with the rest phonons, generating an anti-Stokes signal near centre of the PSF. By filtering out the signal induced by doughnut beam, the effective PSF is hence decreased, meaning a higher spatial resolution. And the resolution can be reduced below 100 nm by properly modifying the probe and additional probe beam. These characters make CARS microscopy especially suit for studying the process of metabolism in live cells.

In the latest 10 years, infrared (IR) microscope has been widely used in many industries, such as material analysis and cellular biology, for its ability of nondestructive imaging and molecular location based on vibrational spectroscopy [26,27]. Traditional IR microscopes are based on either thermal IR sources [28] with low brightness and coherence, such as a globar or Hg-lamp, or synchrotron radiation [29,30]. The synchrotron radiation is complex, expensive, power consuming and always accompanied with intensity fluctuation in spite of high brightness and coherence. The application of non-silica fibers [31-33] brings about a new IR source with both high brightness and coherence, not to mention the broad bandwidth.

In this chapter, the generalized nonlinear Schrödinger equation is briefly deduced, based on the Maxwell's equations. It takes not only the higher-order dispersion into consideration, but also the higher-order nonlinear effects including self-steepening, optical shock formation and intrapulse Raman scattering. Then a predictor-corrector split-step Fourier method is used to simulate the pulse evolutions in both temporal and spectral domain for its high accuracy and fast calculating speed. Based on the fiber structure, the dispersion parameters and nonlinear coefficient can be calculated with finite element method. Then the process of continuum generation in a PCF is described in detail, including pulse evolutions in both temporal and spectral domain. The generated SC has important applications in nonlinear optics, especially CARS microscopy and IR microscopy. Based on the vibrational spectroscopy, CARS microscopy and IR microscopy has been widely used in cellular biology for their label-free and nondestructive imaging. In terms of the limited spatial resolution in CARS microscopy, an APIPD is proposed to break the diffraction barrier, leading to nano-scaled imaging. Furthermore, the combination of this CARS nanoscopy and the broadband continuum is of crucial importance to study the fine structures and metabolic dynamics in live cells. Besides, based on the atomic force microscope, FTIR microscopy has realized a spatial resolution below 100nm. Serving a mid-infrared supercontinuum generated with the chalcogenide PCF as pumping source, this nano-FTIR will become a powerful tool for chemical identification of unknown nanostructures.

2. Numerical modeling of supercontinuum generation

2.1. Nonlinear propagation equation

The propagation equation describing evolution of laser pulses in optical fibers could be derived from Maxwell's equations [34]

$$\nabla \times \mathbf{E} = -\frac{\partial \mathbf{B}}{\partial t}, \quad \nabla \times \mathbf{H} = \mathbf{J} + \frac{\partial \mathbf{D}}{\partial t}, \quad \nabla \bullet \mathbf{D} = \varrho_f, \quad \nabla \bullet \mathbf{B} = 0 \tag{1}$$

where **E** and **B** are electric field vector and magnetic flux density, **H** and **D** are magnetic field vector and electric flux density respectively. **J** and ϱ_f are current density vector and free charge density. They could be related to themselves through the constitutive relations given by

$$\mathbf{D} = \varepsilon_0 \mathbf{E} + \mathbf{P}, \quad \mathbf{B} = \mu_0 \mathbf{H} + \mathbf{M} \tag{2}$$

where ε_0 and μ_0 are vacuum permittivity and permeability, **P** and **M** are the induced electric and magnetic polarizations respectively. In optical fibers, **J**, ϱ_f and **M** all equal to zero, for they are nonmagnetic and absent of free charges.

Based on the Maxwell's equations and constitutive relations, wave equation describing light propagation in optical fibers will be

$$\nabla^2 \mathbf{E} - \frac{1}{c^2}\frac{\partial^2 \mathbf{E}}{\partial t^2} = \mu_0 \frac{\partial^2 \mathbf{P}}{\partial t^2} \tag{3}$$

The induced electric polarization $\mathbf{P}$, which in essence expresses the interaction between light and medium, will depend on the electric field strength as

$$\mathbf{P} = \varepsilon_0 [\chi^{(1)} \bullet \mathbf{E} + \chi^{(2)} : \mathbf{EE} + \chi^{(3)} \vdots \mathbf{EEE} + \ldots] \tag{4}$$

where $\chi^{(k)}(k=1,2,3,\ldots)$ is the k-order susceptibility. The linear susceptibility $\chi^{(1)}$ often contributes to the refractive index and fiber loss, while the second order susceptibility $\chi^{(2)}$ are related to the second order nonlinearity effects such as second harmonic generation (SHG), sum frequency generation (SFG). The third order susceptibility $\chi^{(3)}$ is ofter accompanied by third harmonic generation (THG), four wave mixing (FWM) and so on. The second order susceptibility will be zero, because the fiber is Circular symmetric. So the induced electric polarization is

$$\mathbf{P} = \varepsilon_0 \left[\chi^{(1)} \bullet \mathbf{E} + \chi^{(3)} \vdots \mathbf{EEE} + \ldots \right] \tag{5}$$

Also the induced electric polarization can be divided into two parts, the linear component $\mathbf{P}_L$ and nonlinear component $\mathbf{P}_{NL}$ given by $\mathrm{P_L} = \varepsilon_0 \chi^{(1)} \bullet \mathbf{E}$, $\mathrm{P_{NL}} = \chi^{(3)} \vdots \mathbf{EEE}$. The relations between linear component, nonlinear component and electric field could expressed as

$$\mathrm{P_L} = \varepsilon_0 \int_{-\infty}^{t} \chi^{(1)}(t - t') \bullet \mathbf{E}(r, t') dt' \tag{6}$$

$$\mathbf{P}_{NL} = \varepsilon_0 \int_{-\infty}^{t} dt_1 \int_{-\infty}^{t} dt_2 \int_{-\infty}^{t} dt_3 \int_{-\infty}^{t} \chi^{(3)}(t - t_1, t - t_2, t - t_3) \vdots \mathbf{E}(r, t_1)\mathbf{E}(r, t_2)\mathbf{E}(r, t_3) \tag{7}$$

So Eq. (3) can also be

$$\nabla^2 \mathbf{E} - \frac{1}{c^2} \frac{\partial^2 \mathbf{E}}{\partial t^2} = \mu_0 \frac{\partial^2 \mathbf{P}_L}{\partial t^2} + \mu_0 \frac{\partial^2 \mathbf{P}_{NL}}{\partial t^2} \tag{8}$$

Before solving Eq. (8), some assumptions should be made to simplify it. First $\mathbf{P}_{NL}$ will be treated as a small perturbation to $\mathbf{P}_L$, because the nonlinear changes in the refractive index is very small, less than 10^{-6} in fact. Then the input pulse will keep its polarization when it travels in an optical fiber, to ensure the scalar approximation reasonable. At last, the input field is treated as a quasi-monochromatic light, which is justified when the pulse duration is more than 100 fs.

Supposing the third order susceptibility $\chi^{(3)}$ has the following form

$$\chi^{(3)}(t - t_1, t - t_2, t - t_3) = \chi^{(3)} R(t - t_1)\delta(t - t_2)\delta(t - t_3) \tag{9}$$

with $R(t)$ being the nonlinear resoponse function that has been normalized. So the nonlinear induced electric polarization is

$$\mathbf{P}_{NL} = \varepsilon_0 \chi^{(3)} \mathbf{E}(r, t) \int_{-\infty}^{t} R(t - t_1) |\mathbf{E}(r, t_1)|^2 dt_1 \tag{10}$$

Considering above equations, the numerical equation describing evolution of an optical pulse when it travels in a single-mode fiber is

$$\frac{\partial A}{\partial z} + \frac{\alpha A}{2} - \sum_{k \geq 2} \frac{i^{k+1}}{k!} \beta_k \frac{\partial^k A}{\partial T^k} = i\gamma\left(1 + i\tau_{shock} \frac{\partial}{\partial T}\right)\left[A \int_{-\infty}^{+\infty} R(T') \times |A(T - T')|^2 dt' \right] \tag{11}$$

This equation is known as the generalized nonlinear Schrödinger equation (NLSE). In Eq. (11), the left side models linear propagation effects, with A being the pulse envelope variation in a retarded time frame, α the fiber loss and β_k the k-th order Taylor series expansion of mode propagation constant around the center frequency ω_0 as

$$\beta(\omega)=\beta_0+\beta_1(\omega-\omega_0)+\frac{1}{2}\beta_2(\omega-\omega_0)^2+... \tag{12}$$

The right side of Eq. (11) describes the nonlinear effects, such as self-steepening, Raman scattering, etc. with γ being the nonlinear coefficient expressed as $\gamma(\omega_0)=n_2(\omega_0)\omega_0/cA_{eff}$, where n_2 is nonlinear refractive index, c is the speed of light in vacuum, A_{eff} is effective mode area defined as $A_{eff}=(\iint|F(x,y)|^2dxdy)^2/\iint|F(x,y)|^4dxdy$. Here, $F(x,y)$ is the modal distribution for the fundamental fiber mode. The term of τ_{shock} derives from the frequency dependence of the effective mode area A_{eff} and effective index n_{eff}, associated with effects such as self – steepening and optical shock formation. The relation of the time derivative term τ_{shock} and effective mode area A_{eff}, effective index n_{eff} can be expressed as [35,36]

$$\tau_{shock}=\frac{1}{\omega_0}-\left[\frac{1}{n_{eff}(\omega)}\frac{\partial}{\partial\omega}n_{eff}(\omega)\right]_{\omega=\omega_0}-\left[\frac{1}{A_{eff}(\omega)}\frac{\partial}{\partial\omega}A_{eff}(\omega)\right]_{\omega=\omega_0} \tag{13}$$

In Eq. (13), τ_{shock} could approximately equal to $1/\omega_0$ when the spectral broadening is limited to 20 THz. While the spectral broadening increases to 100 THz or more, the second and third term should be taken into account [37].

R(t) is the nonlinear response function with both the electronic and nuclear contributions included. Since the electronic contribution is nearly instantaneous comparing to the nuclear contribution, R(t) could be written as [38-40]

$$R(t)=(1-f_R)\delta(t-t_e)+f_Rh_R(t) \tag{14}$$

with f_R the fractional contribution of the delayed Raman response to nonlinear polarization, t_e the short delay in electronic response, $h_R(t)$ the Raman response function. Often t_e is very shot (<1 fs) that it could be neglected. So the nonlinear response function will be $R(t)=(1-f_R)\delta(t)+f_Rh_R(t)$. The Raman response function has different forms for fibers made of different materials. For silica fibers, a useful form of Raman response function will be [35]

$$h_R(t)=(\tau_1^2+\tau_2^2)\exp\left(-\frac{t}{\tau_2}\right)sin\left(\frac{t}{\tau_1}\right)\Theta(t)/(\tau_1\tau_1^2) \tag{15}$$

where τ_1 and τ_2 are two adjustable parameters, $\Theta(t)$ is Heaviside function. In a research in 1989, values of τ_1=12.2 fs and τ_1=32 fs were used to model the profile of Raman response, and the result shows good fit to the actual Raman-gain spectrum of silica fiber [35]. Based the peak Raman gain, the Raman response factor could be calculated, saying about 0.18 [38].

In Eq. (15), single Lorentzian profile was used to approximate the actual Raman gain spectrum which is widely used in investigating ultrafast nonlinear process in optical fibers for its simplicity. However, this model underestimates Raman gain below 10THz, while overestimating it beyond 15THz, making the Raman induced frequency shift not so accurate. To address this issue, another model of approximating nonlinear response function was proposed as [41]

$$R(t)=(1-f_R)\delta(t)+f_R[(f_a+f_c)h_a(t)+f_b h_b(t)] \tag{16}$$

with f_a=0.75, f_b=0.21, f_c=0.04, where $h_a(t)$ and $h_b(t)$ have the following forms

$$h_a(t)=(\tau_1^2+\tau_2^2)\exp\left(-\frac{t}{\tau_2}\right)sin\left(\frac{t}{\tau_1}\right)/(\tau_1\tau_1^2) \tag{17}$$

$$h_b(t)=\frac{2\tau_3-t}{\tau_3^2}\exp\left(-\frac{t}{\tau_3}\right) \tag{18}$$

when the parameters are τ_1=12.2 fs, τ_2=32 fs, τ_3=96 fs and f_R=0.245, the simulated Raman gain curve matchs the actual Raman gain profile very well in the whole range of 0-15 THz. To make the investigation more approximate to actual Raman response, a more accurate model could be used although it has more complicated forms [42].

2.2. Frequency-resolved optical gating

In this chapter, we will use the predictor-corrector SSFM proposed by Lee et.al [43] to simulate evolutions of pulses in time and frequency domain. This method shows both fast calculation speed and accurate numerical results. Also the cross-correlation frequency-resolved optical gating method [44-46] (XFROG) is employed to characterize the intensity and phase profiles of the generated continuum. The method describes the pulse structure simultaneously in the time and frequency domain which is intuitive for understanding the dynamics in continuum generation.

In measurement of a continuum based on XFROG, the continuum pulse and a reference pulse should be focused on a nonlinear crystal to generate a sum-frequency signal (SFG signal) after being synchronized in the time domain by a time delay system as shown in figure 1(b). The generated SFG signals under different delay times will be recorded by a spectrometer, with which the XFROG trace could be retrieved. The XFROG trace is

$$I_{XFORG}(\omega,\tau)=\left|\int_{-\infty}^{+\infty}E_{sig}(t,\tau)exp(-i\omega t)dt\right|^2 \tag{19}$$

with the SFG signal field $E_{sig}(t, \tau)=E(T)R(T - \tau)$ where $E(T)$ is the optical field of a continuum which needs to be measured, $R(T - \tau)$ the optical field of reference pulse who has a delay time τ compared to the reference pulse.

It is critical to measure the reference pulse first, because its intensity and phase information will be used to retrieve the continuum pulse. For measurement of a continuum generated in a photonic crystal fiber, the pumping pulse is usually selected to be the reference pulse, although the reference pulse could be any pulse. In measurement of the reference pulse, it will be split into two beams, one as a new pulse to be measured and another as a new reference pulse as shown in figure 1(a). The two beams are focused on the nonlinear crystal to generate a SFG signal that will be measured by the spectrometer. So the FROG trace will be

$$I_{FORG}(\omega, \tau)=\left| \int_{-\infty}^{+\infty} E(T)E(T - \tau)exp(-i\omega t)dt \right|^2 \tag{20}$$

Since the measuring pulse and reference pulse have the same intensity and phase profiles, this FROG is often called self-correlation FROG [47,48].

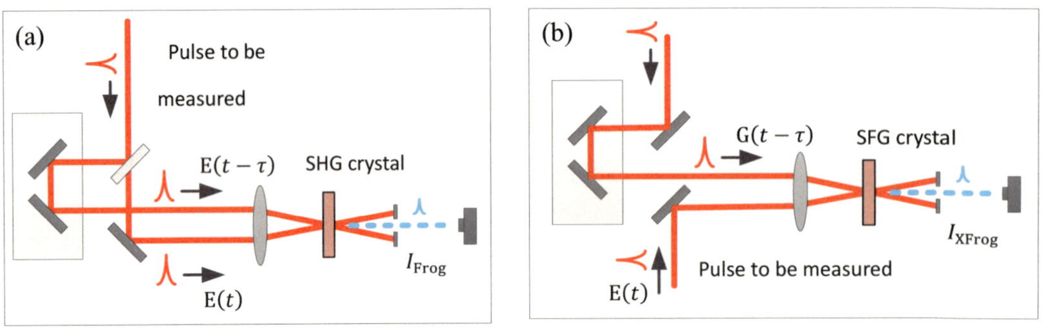

Figure 1. Schematic of (a) auto-correlation FROG and (b) cross-correlation FROG

2.3. Basic numerical results

2.3.1. Calculation of dispersive parameters

The bound electrons in medium act as forced vibration when an optical wave passes through the medium. This interaction often manifests as variation of refraction index, in which the real parts describe the dispersion characteristic and the imaginary parts accounts for absorption of the input wave. When the frequency of input field is away from the resonant frequency, the refraction index could well approximate by the Sellmeier equation [49]

$$n^2(\omega)=1 + \sum_{j=1}^{m} \frac{B_j\omega_j^2}{\omega_j^2 - \omega^2} \tag{21}$$

where ω_j and B_j are the j-th resonant frequency and strength. The parameters ω_j and B_j will change with different core constituents [50]. For bulk-fused silica, the following values of ω_j and B_j are used to fitting the real dispersion curves [51]

$$B_1=0.6961663, \quad \lambda_1=0.0684043\ \mu m$$
$$B_2=0.4079426, \quad \lambda_2=0.1162414\ \mu m$$
$$B_3=0.8974794, \quad \lambda_3=9.8961610\ \mu m$$

The variation of refractive index and dispersion parameter for fused silica with wavelengths are shown in figure 2 with n being the refractive index and n_g being the group index. Group index n_g decreases with wavelength below about 1.31 μm and increases beyond that wavelength point. For this reason, the region below 1.31 μm is often called normally dispersive region, with the other one being anomalously dispersive region. Also the demarcation point between normal and anomalous regimes is known as zero-dispersion wavelength (ZDW).

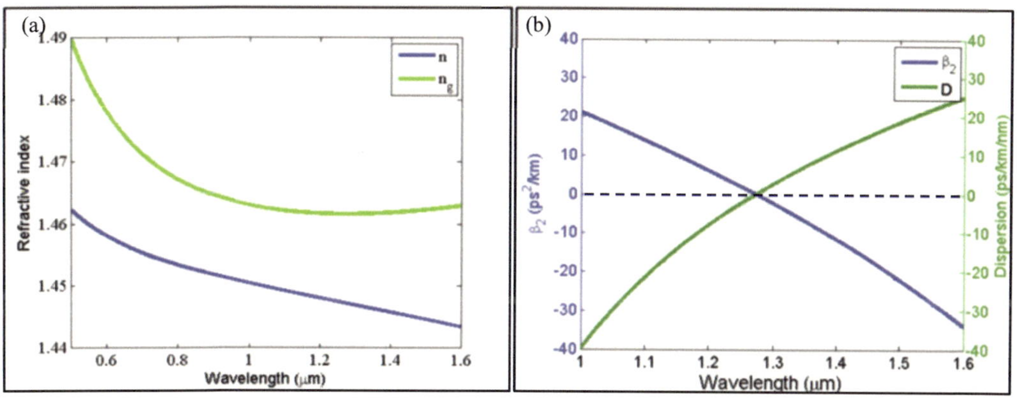

Figure 2. (a) variation of refractive index n and group index n_g with wavelength for fused silica and (b) variation of group-velocity dispersion β_2 and dispersion parameter D with wavelength in bulk fused silica

The dispersive effect in optical fibers is related to the Taylor series of mode propagation constant around the center frequency ω_0

$$\beta(\omega)=n(\omega)\frac{\omega}{c}=\beta_0+\beta_1(\omega-\omega_0)+\frac{1}{2}\beta_2(\omega-\omega_0)^2+\ldots \tag{22}$$

where c is velocity of light, β_k is k-th dispersion coefficient expressed by

$$\beta_k=\left(\frac{\partial^k\beta(\omega)}{\partial\omega^k}\right)_{\omega=\omega_0} \tag{23}$$

The second dispersion coefficient β_2 accounts for the group velocity dispersion (GVD), mainly leading to pulse broadening. In fact, another dispersion parameter D is usually used to describe the dispersive effect.

$$D = -\frac{2\pi c}{\lambda^2}\beta_2 = -\frac{\lambda}{c}\frac{\partial^2 n}{\partial \lambda^2} \tag{24}$$

Dispersive effect in a PCF is associated with both the material dispersion and waveguide dispersion while it mainly arises from material dispersion in bulk materials. Therefore, in terms of dispersion in a PCF, the fiber structure and index distribution should be both taken into account. Considering a solid-core PCF made of fused silica, the finite element method (FEM) is used to model the fundamental mode distribution for different input wavelengths [52-54]. In the modeling, the air hole diameter d and pitch Λ are set to be 1.4 μm and 1.6 μm respectively [6], leading to the cross-section profile of the PCF as shown in figure 3(a). Figure 3(b) shows a two-dimension distribution of the fundamental mode for input wavelength λ=0.4 μm with the material dispersion included. Based on this simulation, the effective index of the PCF n_{eff} is about 1.462353.

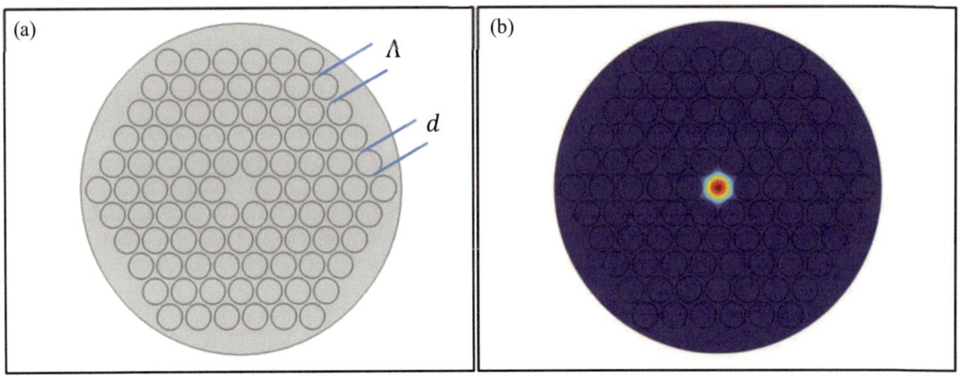

Figure 3. (a) Cross-section model of the PCF; (b) Two-dimension distribution of the fundamental mode

By calculating every n_{eff} corresponding to different input wavelength, the variation of effective index with wavelength is fitted using MATLAB. Then the GVD parameter β_2 and dispersion parameter D are obtained with Eq. (24) and the profiles with wavelength are plotted in figure 4 respectively. Obviously the ZDW locates at about 730 nm and there is only one ZDW in the range from visible light to near-infrared.

2.3.2. *SC generation with PCF*

In this section, we numerically simulate the SC generation process in the PCF described above based on NLSE in Eq. (11) by PC-SSFM. In the simulation, the more accurate expression for nonlinear response function as in Eq. (18) is used. We first consider the SC generation with

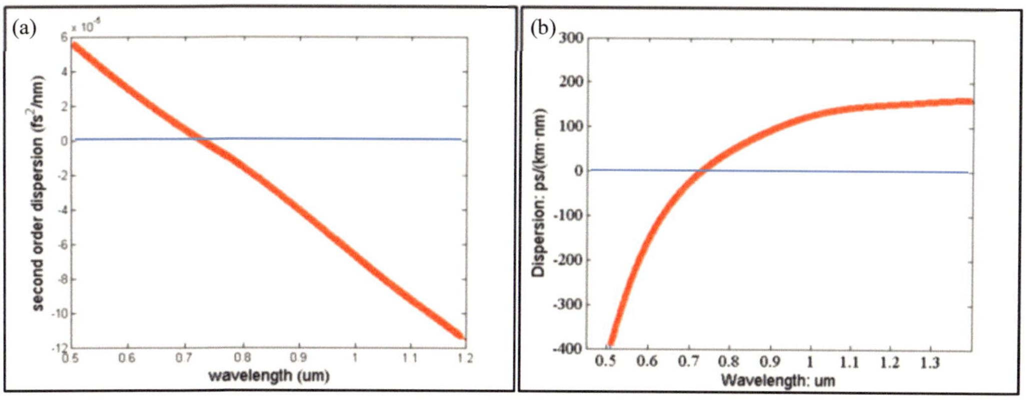

Figure 4. Variation of GVD parameter β_2 (a) and dispersion parameter D (b) with wavelength

PCF described above with hole diameter 1.4 μm and hole pitch 1.6 μm. Because of the ZDW about 730 nm, pump pulses with central wavelength of 800 nm are used to launch the fiber where the nonlinear parameter is estimated to be 0.113 $W^{-1}m^{-1}$. And the Taylor series expansion coefficients are $\beta_2 = -1.581 \times 10^{-5} fs^2 / nm$, $\beta_3 = 7.819 \times 10^{-5} fs^3 / nm$, $\beta_4 = -1.015 \times 10^{-4} fs^4 / nm$, $\beta_5 = 2.549 \times 10^{-4} fs^5 / nm$, $\beta_6 = -4.808 \times 10^{-6} fs^6 / nm$.

Assume the initial input pulse has a hyperbolic scant profile that could be expressed as

$$A(0, t) = \sqrt{P} \operatorname{sech}\left(\frac{t}{t_0}\right) \exp\left(-\frac{iCt^2}{2t_0^2}\right) \tag{25}$$

with$P$ being the peak power, t_0 the pulse duration at $1/e$-intensity point, C the initial chirp parameter. It is worth noting that the full width at half maximum (FWHM) is $2\ln(1+\sqrt{2})$ times as large as t_0. In this simulation, the peak power and pulse duration are set as $P = 10$ kW , $t_{FWHM} = 50$ fs, neglecting the initial chirp.

The temporal and spectral evolutions of input pulse with propagation distance are depicted in figure 5. In the initial stage of propagation, the spectral broadening is almost symmetric. And the broadening mainly occurs within this first 6 mm, during which the pulse is compressed strongly. After that, the spectral broadening becomes strong asymmetric, with distinct peaks on the long and short wavelength sides. The long wavelength components manifest the breakup of input pulse into several sub-pulses, known as soliton fission [52] caused by high-order dispersion, self-steepening effect and Raman scattering. The short wavelength components are related to SPM effect and dispersion wave generation [55,56].

With longer propagation, the SC shows larger spectral broadening mainly on the long wavelength side due to soliton self-frequency shift (SSFS) induced by stimulated Raman scattering, while the spectral broadening on the short wavelength side almost remains unchanged. The spectral broadening process is often accompanied with re-distribution of

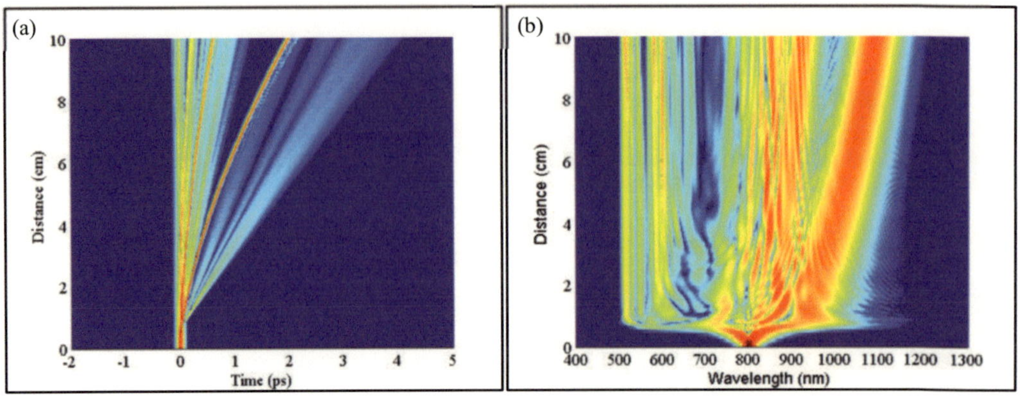

Figure 5. Temporal (left) and spectral (right) evolution over a propagation distance for input pulse centered at 800 nm with peak power 10kW and pulse duration 50fs

energy, that the long wavelength components possess more while the short wavelength components possess less. This is explained by that: With the blue-shift components as pumping light, Raman gain will amplify the red-shift components effectively, leading to energy's transferring from blue components to red components. This energy transfer describes the red shift of the soliton spectrum with propagation distance.

To study the detailed characteristics of the generated SC, temporal and spectral slices at 15 cm propagation distance are plotted in figure 6. The time structure and spectrum are both shown on a linear scale and the intensity is normalized. There are three main peaks in the time domain located at 0.46 ps, 1.2 ps, and 3.8 ps, corresponding to the distinct peaks at 870 nm, 953 nm, and 1143 nm respectively in the frequency domain. These three peaks are caused by soliton fission and their central positions in both time and frequency domain are determined by their intrinsic characteristics, SSFS effect and the propagation distance.

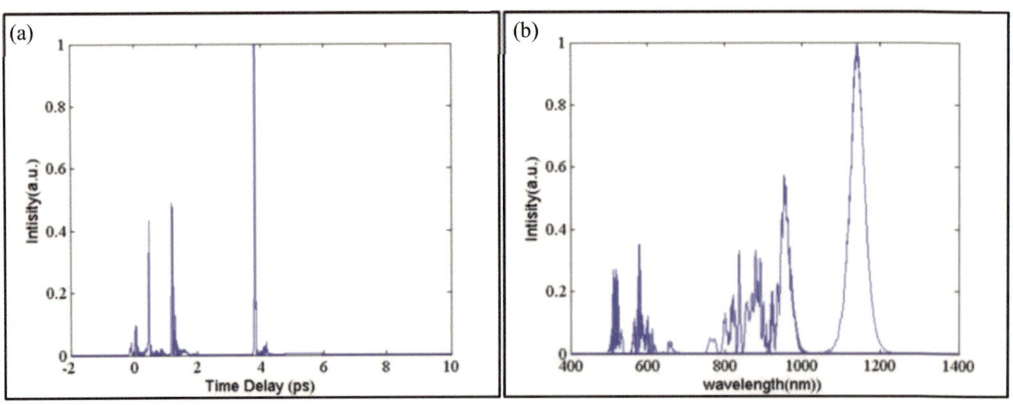

Figure 6. Pulse characteristics in (a) time and (b) frequency domain after 15 cm propagation

To observe the detailed structure in figure 6(a), magnified profiles of the segments ranging from 1.1 ps to 1.8ps, and 4ps to 7ps are plotted in figure 7(a) and (b) respectively. In figure (a), the ultrafast oscillating structure is associated with two sidelobes on both side of input wavelength. The main peak belongs to a second soliton and the oscillating structure results from beating between the two sidelobes. The relation of them is easy to see in figure 7(c) with 370 nm separation in spectrum at delay time 1.2 ps. There are still some fine oscillating structures lower than the two peaks in figure (a) ranging from 1.4 ps to 1.8 ps. There exists a little bump ranging from 4.6 ps to 6.5 ps in figure (a) with figure (b) being the magnification of partial structure. This bump is related to the dispersion wave generation around 546 nm.

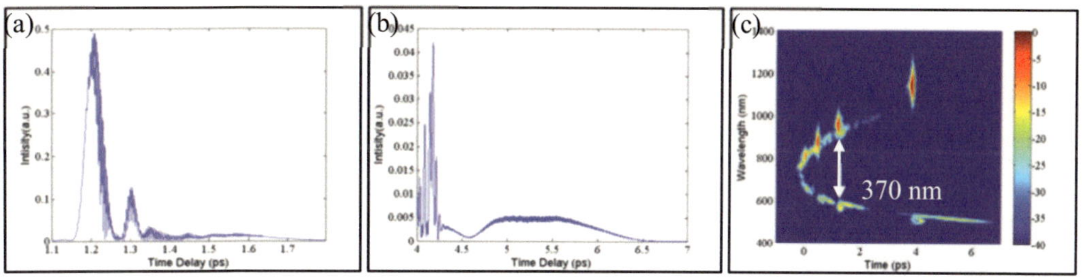

Figure 7. (a) Oscillating structure ranging from 1.2 ps to 1.7ps (b) Generated dispersion wave ranging from 4.6 ps to 6.5 ps (c) Spectrogram with 15 cm propagation distance

3. Supercontinuum used in coherent anti-Stokes Raman scattering nanoscopy

3.1. Theories of CARS process

Coherent anti-stokes Raman scattering (CARS) has proven to be a promising nonlinear optical technique that is capable of obtaining high-sensitivity and three dimensional images based on molecular vibrational spectrum without labeling [57,58]. CARS is an example of a four-photon process. The first two light-matter interactions, one with pump frequency ω_L and another with Stokes frequency ω_S , set to that of a vibrational resonance, a pump photon is annihilated, correspondingly a Stokes photon and a phonon (i.e., the energy quanta of the material excitation)is generated, the number of phonons is equal to the generated Stokes photons. The second light-matter interactions, the energy quanta of the material excitation, phonons, interact with the probe photons to generate a coherent anti-Stokes signal. In general, both processes occur simultaneously and the whole process can be treated as a four-photon process [59]. The detected signal strength depends nonlinearly on incident intensity.

The full quantum picture of CARS process is depicted in Fig.8. $|e$ and $|g$ correspond to the vibration excited and ground state respectively. For simplicity of analysis, a single frequency model is provided. All laser fields-the pump, Stokes, probe, and phonons (or molecules) are quantized. The whole CARS process can be divided into two individual steps as shown in Fig.

8. In the first step, a coherent phonon population is directly created by coherent Raman scattering when a pump beam at frequency ω_L and a Stokes beam at frequency ω_S with frequency difference $\omega_L - \omega_S$ tuned to a Raman-active molecular vibration ω_v arrive at the sample simultaneously via a high numerical aperture objective. The process is impulsively stimulated Raman scattering (ISRS), which is considered as the main mechanism of coherent phonon generation [60,61]. The coherent phonons are bosons and do not obey an exclusion principle. They are in a non-equilibrium state and in phase. In the second step, the coherent phonons interact with a time-delayed probe beam at frequency ω_P to generate a blue-shift anti-Stokes signal at $\omega_{AS} = \omega_P + \omega_v$, leading to separation of the CARS signal from the incident laser beams conveniently and efficiently rejecting fluorescence. Delay time between coherent phonons and probe beam is certainly much shorter than the vibrational dephasing time.

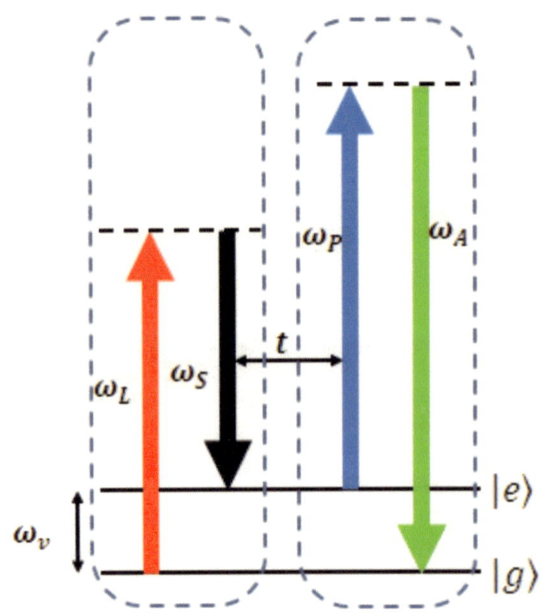

Figure 8. Energy level diagram of the two individual steps in CARS

The quantum mechanical treatment of the first process, which is a second-order process involving two light-matter interactions, resembles the spontaneous Raman effect. The important difference of the two is that in CARS process, the Stokes light stems from an applied laser field. Spontaneous Raman is a weak effect because the spontaneous interaction through the vacuum field occurs only rarely. This weakness can be overcome when the spontaneous nature of the initial state to final state $i \rightarrow f$ transition is eliminated by applying a second field of frequency ω_s. The treatment is based on the calculation of the transition rate between the states of the molecule, which is described by Fermi's golden rule. The equation (26) predicts the growth of the generated Stokes photons, i.e., phonon number [62,63]

$$\frac{dn_S^{inc}}{dl} = N \bullet \frac{\pi e^4 \omega_L \ \omega_S}{2\varepsilon_0^2 \hbar^2 V\ ^2\mu_L^2\ \mu_S c} \mid \alpha_R \mid^2 n_L^* \left[n_S^* + 1\right] \delta\left(\omega_S - \omega_L + \omega_v\right) \tag{26}$$

Consider a volume V , which contains N molecules and has a length l parallel to the beam propagation direction. The Raman transition polarizability α_R plays a role similar to the classical polarizability change $(\partial\alpha / \partial q)$. ω_L and ω_S is the frequencies of pump and Stokes light, respectively. e is the electron charge, ε_0 is the vacuum permittivity, $\hbar$ is Planck's constant, and δ is the Dirac delta function.μ_L and μ_S denote the refractive indexes. n_S^{inc} indicates the increased (decreased) number of Stokes (pump) photons, i.e., the phonon number of energy $\hbar\omega_v$, n_{phonon}. In the CARS process, the Stokes light stems from an applied laser field, and the n_S^* consists of the applied Stokes photon n_S and increased number n_S^{inc} that stems from the first two light-matter interaction process. Similarly, n_L^* , amounts to the pump photon number n_L subtracting the decreased number n_S^{inc}. With the use of the pump and Stokes laser light sources of sufficiently high power, $n_S^{inc} \ll n_L$, n_S ,the present pump and Stokes number $n_L^* = n_L - n_S^{inc}$, $n_S^* = n_S + n_S^{inc}$, approximately amount to n_L and n_S, and the factor 1 in equation (26) can be omitted at the same time. The laser intensities I_L and I_S exist on the focal plane in the form of Gaussian distribution, and we adopt an infinitesimal region in the focal plane where the intensities can be supposed constant $I_L = (\hbar\,\omega_L\, c / \mu_L\, V) n_L$, $I_S = (\hbar\,\omega_S c / \mu_S V) n_S$. In an fs-CARS transient, the pump and Stokes beams temporally overlapped, coherently excite the molecular field, the duration of the transitory process denoted by the coherent excitation time is negligible. The homogeneous broadening of the atomic final state converts the delta function to a Lorentzian lineshape. when $\omega_S = \omega_L - \omega_v$, $\delta\left(\omega_S - \omega_L + \omega_v\right) \rightarrow \frac{\Gamma / \pi}{(\omega_S - \omega_L + \omega_v)^2 + \Gamma^2} = \frac{1}{\pi\Gamma}$, where Γ denotes the linewidth relating to the dephasing time T_2 of the physical system, $\Gamma = 1 / T_2$.

Accordingly, from the equation (26), we obtain:

$$n_S^{inc} = \frac{I_S \mu_S V}{\hbar\omega_S\, c} \bullet I_L\, gl \tag{27}$$

where g is the gain coefficient, N_0 is the molecule density.

$$g = N_0 \bullet \frac{e^4}{2\varepsilon_0^2 \hbar^3 V\ ^2\mu_L^2\ \mu_S c^2} \frac{\omega_S}{\mu_L\ \mu_S} \mid \alpha_R \mid^2 \frac{1}{\Gamma} \tag{28}$$

Then, considering a molecule with just a single vibrational transition, the coherent phonon number per time unit in the first ISRS step can be written as

$$n_v = \frac{I_S \mu_S V}{\hbar\omega_S\, c} \bullet I_L\, gl \tag{29}$$

where n_v indicates the coherent phonon number. The interaction volume V, which contains N molecules, is defined by the diffraction-limited focal spot area A_{foc} and the interaction length l paralleling to the beam propagation direction. For $g \propto N_0$, it is apparent that the coherent phonon number is linearly dependent on N_0.

The phonon number subjects to a saturation effect and will not increase infinitely with the growth of incident light powers [64]. The maximum of coherent phonon number per pulse is molecule number in the focal volume. In another word, all of the molecules in the focus are excited to the vibrational state.

$$n_v^{pulse} = \frac{I_S \mu_S V}{\hbar \omega_S c} \bullet I_L \, g\tau \leq N \tag{30}$$

Here, n_v^{pulse} is the coherent phonon number per pulse. The magnitude of the spot is roughly estimated as $\sim 10^{-9}\ \text{cm}^2$, corresponding to a focal volume $V \sim 8 \times 10^{-14}\ \text{cm}^3$. The gain coefficient is $g \sim N_0 \times 10^{-31} \text{cm}/\text{W}$. From equation (30), we see that saturation of coherent phonons results in product of the pump and Stokes intensities not exceeding $\sim 10^{18} \text{W}^2/\text{cm}^4$. Therefore in the first step of a T-CARS process, the pump and Stokes power density at focus is on the same order of $\sim 10^9 \text{W}/\text{cm}^2$. In principle, if the pump and Stokes intensity exceed such an extreme value, the further increased intensity will not contribute to generation of the coherent phonons any more. In practice, however, energy of the pump and Stokes may be higher than that theoretical estimation due to the spectral broadening of femtosecond laser pulses.

Hereafter, we analyze the second step in CARS process. In the spontaneous Raman anti-Stokes process, the involved lattice vibrations called incoherent phonons are thermally excited with a density function n_v given by the Bose-Einstein distribution. In this case, oscillatory phases of the phonons are completely random. Thus, individual oscillatory motions are canceled out by each other. Spontaneous Raman scattering is an incoherent linear process and the incoherent signal is randomly scattered to 4π solid angle, then the collected signal is [65]

$$P_A = N \left(\frac{\partial \sigma}{\partial \Omega} \right)_A n_v I_L \, \Omega_{collection} \tag{31}$$

with P_A representing the anti-Stokes power, I_L the incident laser intensity and $\Omega_{collection}$ the effective collection angle. $\left(\frac{\partial \sigma}{\partial \Omega} \right)_A$ is the spontaneous Raman differential scattering cross section. The scattered signal is proportional to the incident laser intensity and the occupation number of the upper Raman level. Different from spontaneous anti-Stokes Raman scattering, the second step in CARS is via introducing probe photons to interact with the coherent phonons, which are in a non-equilibrium state and in phase. Delay time between coherent phonons and probe beam is certainly much shorter than the vibrational dephasing time. Therefore, relaxation of the phonons is omitted, that is, all phonons are considered to have chances to interact with the probe photons and the impact of the delay time on the signal strength can be readily omitted. An important advantage of CARS microscopy over spontaneous Raman scattering microscopy is the fact that CARS signal is emitted into a $10^{-4}sr$ range, comparing to spontaneous Raman signal randomly scattered into 4π solid angle, which provides an important information that scattering cross section $\partial\sigma_C$ for the interaction of coherent phonons and probe photons is estimated to be enhanced by about five orders of magnitude [66]. In some sense, coherent process is equivalent to increasing the scattering cross section. The coherent anti-Stokes photons are confined to such a small solid angle, thus a condenser can almost completely collect the forward propagating signal. For low concentration samples, the distance between adjacent solute molecules is so great that energy of the coherent phonons cannot effectively transmit from one molecule to another. Accordingly, in the second step, the anti-Stokes signal strength is exclusively proportional to the number of coherent phonons and then linear dependence on the concentration of solute molecules. Comparing with incoherent (spherical) spontaneous Raman process, we calculate the number of the detected photons for coherent (directional) CARS in the repetition mode as

$$n_{CARS}^{low} = n_v^{pulse} \bullet n_P \bullet \frac{1}{A_{foc}} \partial\sigma_C \bullet \Delta t f_{rep} \tag{32}$$

In contrast, for high concentration samples, the coherent phonon energy can freely transmit among solute molecules in the focal volume, therefore in the second step, anti-Stokes signal strength isn't simply proportional to the coherent phonon number but also concerns with the molecule number. The overall T-CARS process depends quadratically on concentration, hence the anti-Stokes photon number in the repetition mode is

$$n_{CARS}^{high} = N \bullet n_v^{pulse} \bullet n_P \bullet \frac{1}{A_{foc}} \partial\sigma_C \bullet \Delta t f_{rep} \tag{33}$$

where, n_{CARS}^{low} and n_{CARS}^{high} denote anti-Stokes photon number in the low and high concentration samples, respectively. n_P is photon number of the probe beam per pulse. The three pulsed near-infrared excitation beams are tightly focused with a 1.2NA water-dipping lens for deeper penetration depth to the spot area A_{foc}. Δt is exposure time. N is molecule number in the foci. In short, for $n_v^{pulse} \propto N_0$, the anti-Stokes signal strength is strictly linear and quadratic on concentration in the low and high concentration samples, respectively, which conflicts with the popular statements of CARS signal consistently quadratic concentration dependence. The linear concentration dependence is especially beneficial to quantitative analysis for low concentration samples [67-70]. In the cross area between the low and high concentration, the observed concentration dependence from the relative experiments is neither quadratic nor linear, i.e., $n_{CARS} \propto N^{\alpha}$, where $1<\alpha<2$. An experimental proof in Figure 9 shows the dependence of T-CARS signal strength on the concentration of alcohol in pure water at C-H stretch band [66,71] (2900cm^{-1}). We record the signal intensity at every volume fraction by 20 times to exclude the influence of signal fluctuation. When volume fraction of alcohol gradually increases, concentration dependence transits from strictly linear to sub-quadratic, until quadratic. It is apparent that when volume fraction of alcohol in water is lower than 20%, the concentration dependence is strictly linear. We prepare the aqueous solution of alcohol with smaller volume fraction intervals to obtain a more precise fitting result. This experiment result consists with the theoretical analysis, and more different samples will be prepared to verify the universality of the theoretical analysis.

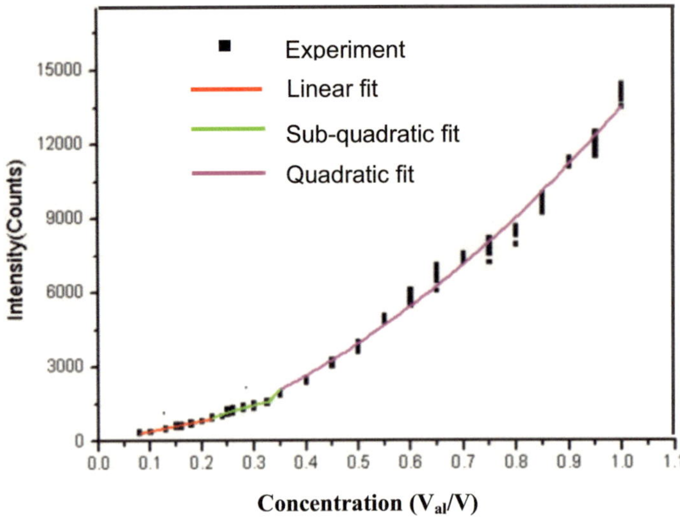

Figure 9. T-CARS intensity for increasing volume fractions of alcohol in water at different volume fraction from 7% to 100%.

3.2. Break the lateral diffraction barrier in CARS microscopy

The application of CARS microscopy is limited, nevertheless, by the acquirable lateral spatial resolution, typically around 300nm. With anti-Stokes signal generated from the focal spot, which is then scanned to build up a an image of the sample, a promising way to generate sub-diffraction-limited CARS images is via minimizing the spot to nanometer extent, as in STED [20]. Several schemes have been suggested recently for breaking the diffraction barrier in CARS microscopy, and can be mainly classified as three means [72-74]: by use of an additional saturation laser to populate coupled vibrational levels; a phase-controlled local oscillator generating ring field at the anti-Stokes frequency being out of phase with the induced CARS field in the focal volume; to add structured illumination for achieving widefield CARS microscopy. Whereas, these schemes still rest on the theory, and no experimental implementations have been reported. And the flows of these schemes are obvious: it's difficult to implement the schemes and can only obtain resolution enhanced images of the molecule specified by the anti-Stokes Raman spectral signal for its particular chemical-bond, not its broadband even total CARS spectral signals; the second method must know the type of the bond in advance, and the third can merely obtain a resolution of around 120nm, etc.

We suggest an approach, the so-called additional probe-beam-induced phonon depletion (APIPD) method [25], to visualize the resolution enhancement by introducing an additional doughnut probe beam with a wavelength that is different from the Gaussian probe beam, which almost synchronizes with the pump and Stokes beams, to deplete the phonons at the periphery of the focal spot. The Gaussian probe beam that immediately follows, with a delay time less than 1ps, will yield anti-Stokes signal in the center but not at the periphery of the focal spot. The difference of the two probe beams in wavelength leads to the disparity of two anti-Stokes signals in spectra, and the signals at the periphery can be filter out by virtue of a particular filter. Consequently, the effective focal spot can be substantially reduced. Super-resolution images can be obtained by scanning the suppressed focal spot.

The rationale of our scheme is to suppress the anti-Stokes signal generation at the periphery of the diffraction-limited spot, i.e., deplete the phonons produced by the pump and Stokes light in this region, then no phonon interact with the probe photons to generate CARS signal here anymore. We adopt an additional probe beam, the first one, with frequency ω_{P1}, which is doughnut-mode, analogy to STED, to achieve this by interacting with phonons to generate useless anti-Stokes signal with frequency ω_{A1} at the periphery which can be filtered out through a specific filter. The probe beam (ω_{P1}) is nearly simultaneous with the pump and Stokes beams on the sample. The intensity of the first probe beam (ω_{P1}) is strong enough to completely deplete the phonons at the peripheric region, while the immediately followed second probe light (ω_{P2}) with Gaussian profile only generates the useful anti-Stokes signal (ω_{A2}) at the central part of the spot. The time delay between the two probe beams is certainly much smaller than the vibrational dephasing time. The difference of ω_{P1} and ω_{P2} results in the disparity of ω_{A1} and ω_{A2}, therefore we can select the specific filter to filter out the unwanted anti-Stokes signal of frequency ω_{A1} to obtain the useful signal ω_{P2} at the suppressed extent of the spot. The process of the APIPD is illustrated as in figure 10.

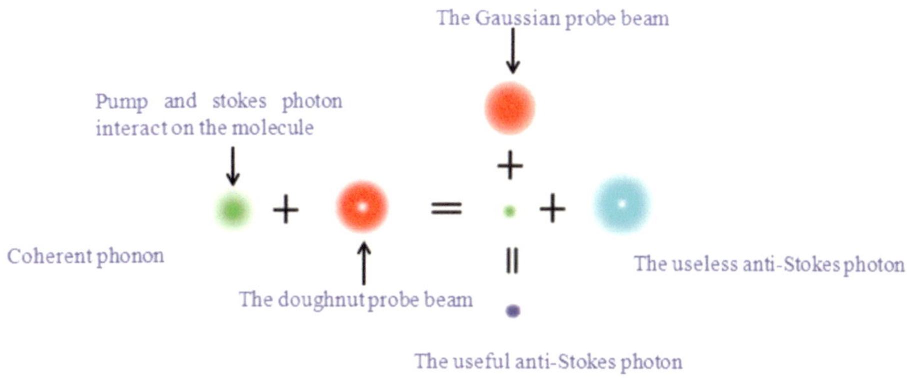

Figure 10. The illustration of the process of the APIPD method

$f_S(r)$ and $f_L(r)$ are the Gaussian Stokes and pump beam distribution which can be replaced by squared cosine function, corresponding to that in equation (34).$f_{P1}(r)$ and $f_{P2}(r)$ are the first Gaussian and second doughnut probe beams, respectively, which can be similarly replaced by the squared sine and squared cosine functions. Introducing the peak intensities of I_L^{max} , I_S^{max} , I_{P1}^{max} and I_{P2}^{max} representing the pump, Stokes, the first and second probe beams respectively, we have four beams' spatial distribution

$$\begin{aligned} f_L(r) &= I_L^{max}\cos^2(\pi r\mu_L \sin\alpha/\lambda_L) \quad (a) \\ f_S(r) &= I_S^{max}\cos^2(\pi r\mu_S \sin\alpha/\lambda_S) \quad (b) \\ f_{P1}(r) &= I_{P1}^{max} sin^2(\pi r\mu_{P1} sin\ \alpha/\lambda_{P1}) \quad (c) \\ f_{P2}(r) &= I_{P2}^{max} cos^2(\pi r\mu_{P2} sin\ \alpha/\lambda_{P2}) \quad (d) \end{aligned} \tag{34}$$

where α is the semi-aperture of lens and μ_L , μ_S , μ_{P1}, μ_{P2} are the refractive indexes, λ_L , λ_S , λ_{P1} , λ_{P2} are the wavelength of the four beams, respectively.

The first doughnut probe light (ω_{P1}) with intensity distribution $f_{P1}(r)$ on the focal plane, almost synchronizes with the pump and Stokes light on the sample. The second Gaussian probe light (ω_{P2}) with intensity distribution $f_{P2}(r)$ follows the first doughnut subsequently within 1 ps much smaller than the vibrational dephasing time. Consequently, after the course of the first probe light, the effective point-spread-function (PSF) of the useful anti-Stokes signal (ω_{A2}) at the suppressed extent of the spot can be expressed as:

$$h^*(r) = f_{P2}(r)\left[n_v - n_v \bullet n_P \bullet \frac{1}{A_{foc}}\partial\sigma_C\right]$$

Hence, the detailed expression of $h(r)$ is

$$h(r) = \begin{cases} h^*(r), & 1 \geq n_P \bullet \frac{1}{A_{foc}}\partial\sigma_C \\ 0, & \text{others} \end{cases} \tag{35}$$

The first term in the square bracket of the equation (35) denotes the total phonon number, and the second term is the phonon number interacting with the first doughnut probe photons at any position of the focal volume. Here A, the cross section area of the volume V, is V/l. N is the molecule number. When the second term in the square bracket is not larger than the first term, i.e., phonon number interacting with the first doughnut probe beam is not larger than the total phonon number generated by the pump and Stokes beams, the expression $h^*(r)$ is tenable, or else, the phonons have been depleted before the second probe beam arrives and $h(r)=0$, therefore, the PSF in the whole space is a piecewise function expressing as:

$$h(r) = \begin{cases} h^*(r) 1 \geq n_P \bullet \frac{1}{A_{foc}}\partial\sigma_C \\ 0 \qquad\qquad\qquad else \end{cases} \tag{36}$$

The full width at half maximum (FWHM) of $h^*(r)$, denoted by Δr, is approximated by a Taylor series expansion of $h^*(r)$ to the second order:

$$\Delta r = \sqrt{2}\frac{\lambda}{\pi n \sin\alpha\sqrt{3+K}} \approx \frac{0.9}{\sqrt{3+K}} \bullet \frac{\lambda}{2n\sin\alpha} \tag{37}$$

Here, influence of the refractive index difference and wavelength difference of the pump, Stokes and probe beams are negligible, and both of them are uniformly expressed as λ and n. K describes magnitude of the doughnut probe beam. Resembling the STED microscopy, the effective PSF, in which the useful anti-Stokes signal (ω_{A2}) is measured, becomes deformed by the unwanted anti-Stokes signal (ω_{A1}) generation. Resolution is enhanced via a factor of square root of the nonlinear order K , namely depletion level, defined by $K = I_{P1}^{max} / I_{P2}^{max}$. Apparently, if no doughnut probe beam introduced, namely $K = 0$, the equation (37) largely reproduces the CARS nonlinear optical microscopy's lateral resolution, which is about $\sqrt{3}$ times better than that of linear optical microscopy. With the gradually increasing of K, the lateral spot width decreases continually, following a square-root law, i.e., resolution is increased by a factor of square root of K. In principle, almost no limitation to super-resolution can be obtained. The resolution enhancement technique can yield an effective focal spot which can be fundamentally reduced, theoretically, to an infinitesimal spot, approximating to the size of a molecule or even further to sub-molecular dimensions. Scanning the suppressed spots automatically renders images with resolution breaking the Abbe's diffraction barrier. Our approach can only fundamentally reduce the focal spot in the lateral direction.

Taking $K \approx 50$ to obtain a lateral resolution offor instance, the lateral resolution attainable using this microscopy is approximately ~40 nm, indicating an approach to 5-fold improvement in x or y direction over the diffraction barrier. The PSFs of the traditional and the resolution enhanced CARS microscopy are displayed in figure 11 exploiting the same parameter.

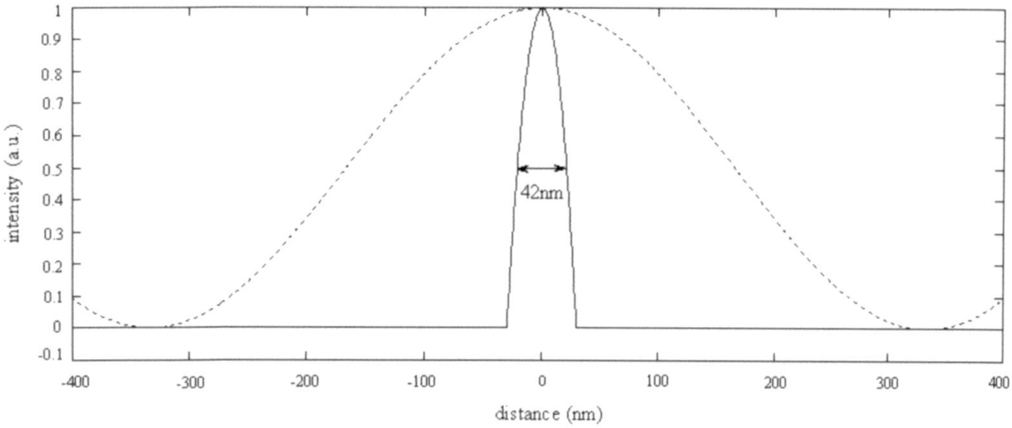

Figure 11. Simulation of the point spread function (PSF) of the proposed resolution enhanced CARS microscopy (green, solid line) contrasting with the traditional system (black, dashed line). The new technique indicates that roughly approximate to 8-fold improvement in lateral direction over the diffraction barrier.

In the APIPD scheme, a super-continuum (SC) source generated from a photonic crystal fiber (PCF), as the pump and Stokes pluses, is provided with broad spectral width that allows simultaneous detection over a wide range of Raman shifts. We adopt 500nm and 800nm as the central wavelength of the first and second probe beam, resulting in absolutely distinguishing the two sets of anti-Stokes signals from each other in spectrum. The unwanted anti-Stokes signal on the periphery can be filter out to achieve resolution enhanced images of the molecule specified by its broadband even total CARS spectral signals not only by the anti-Stokes signal of its particular chemical-bond.

3.3. SC used in CARS nanoscopy

The traditional CARS device consists of a pump and stokes pulses with a frequency ω_L and $\omega_S$respectively. And the two beams focus onto a sample after being temporally overlapped. The two waves interact together through the third-order susceptibility to generate a blue-shift anti-stokes signal. CARS being a resonant process, the anti-Stokes signal is strongly enhanced when $\omega_L - \omega_S = \Omega_r$.To avoid the use of two synchronized laser oscillators emitting the pump and Stokes frequencies required in the process, several previous studies have demonstrated multiplex CARS system by using photonic crystal fibers (PCF) [4]. Super-continuum (SC) source generated from a PCF, as the pump and Stokes pluses, is provided with broad spectral width that allows simultaneous detection over a wide range of Raman shifts. The specific and decisive technical advantage in using PCF for CARS measurements is the possibility to generate both pump and Stokes radiations from a single laser. Figure 12 depicts the simple and multiplex CARS process in an energy diagram, the vibrational (or rotational) levels of energy of molecules being schematized by ω_R.

In our scheme, a continuum pulse generated from a PCF, as the pump and Stokes pluses, is provided with broad spectral width that allows simultaneous detection over a wide range of Raman shifts. We adopt 500nm and 800nm as the central wavelength of the first and second probe beam, resulting in absolutely distinguishing the two sets of anti-Stokes signals from each other in spectrum. The unwanted anti-Stokes signal on the periphery can be filter out to achieve resolution enhanced images of the molecule specified by its broadband even total CARS spectral signals not only by the anti-Stokes signal of its particular chemical-bond.

Good compatibility with practical application requirements and acquisition time in CARS microscopy is made with a continuum pulse serving as the Stokes which is generated by injecting ultrashort pulses into a PCF with high nonlinearity. Because of its broad spectra range, frequency difference between the pump and Stokes can cover most of the biologically interesting fingerprint region. Therefore, most distinct vibration modes, perhaps full modes even, can be probed simultaneously by a probe pulse as long as different frequency components in the generated SC propagate at about a same speed, known as broadband CARS spectroscopy.

It is critical to generate a very SC pulse to meet its application in CARS microspectroscopy. To acquire the full CARS spectrum of a certain molecule, the frequency difference should at least cover the fingerprint region from 500 cm^{-1} to 3000 cm^{-1} at a same moment. This determines that the generated SC pulses have the following characteristics: (i) the spectral bandwidth is broad

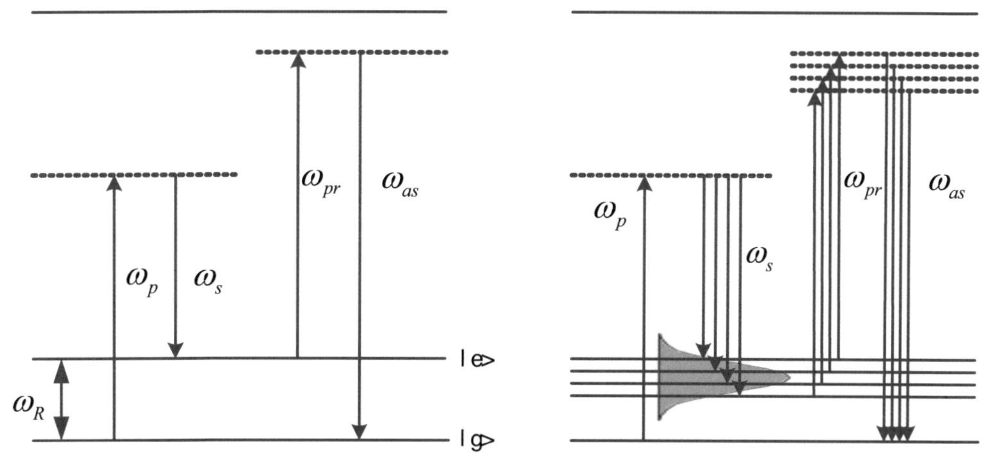

Figure 12. Single CARS process (left) and multiples CARS process with a broadband Stokes pulse

enough, (ii) various components in SC propagate at about a same speed, (iii) spectral intensity should be large and uniform.

3.3.1. SC generation in traditional PCFs

Although dispersive property of a PCF leads to different propagating velocities of different frequency components in the generated SC, soliton pulse trains from a PCF have been considered as the Stokes in many CARS spectroscopy schemes [75-77]. In this part, we will detailed discuss the generation and characteristics of the soliton pulse trains used in CARS spectroscopy.

When an ultrashort pulse travels in a nonlinear material such as a PCF, different frequency components in it will propagate at different velocity, because of the dispersive effect, including material dispersion and waveguide dispersion. In the anomalous regime, this dispersive effect manifests blue-shift components moving faster than the red-shift components, leading to the temporal broadening. At the same time, SPM generates a nonlinear chirp which is negative near the leading edge and positive near the trailing edge. Therefore the propagating speed decreases near the leading edge and increases near the trailing edge, making the pulse narrow down. Under certain conditions, the optical envelope will keep its shape and propagating velocity known as optical soliton [78,79].

As discussed in section two, distinct peaks in time domain are obvious. Because the soliton ejected earlier has larger group velocity, higher peak power and shorter durations, it will propagate faster than others. So the separation between these peaks becomes larger with a longer propagation distance. Figure 13 shows the spectrograms of the generated SC when the propagation is 10 cm, 15 cm, 20 cm and 25 cm. The input pulse is central at 800 nm with peak power 10kW and pulse duration 50fs as before. From the figures, the first-order soliton shifts from 2 ps to about 8.5 ps when the propagation changes from 10 cm to 25 cm gradually. Also the time delay between solitions becomes larger with a longer propagation distance by making contrast of the four graphs. For the dispersion wave on the short wavelength side, time delay between different frequency components also becomes larger. The time span is about 2 ps when the propagation distance is 10 cm, while it is about 5 ps with 25 cm travelling distance. The shifting of solitons in time domain is corresponding to the spectral shifting. The central wavelength of the first-order soliton shifts from 1100 nm to 1200 nm with the propagation from 10 cm to 25 cm.

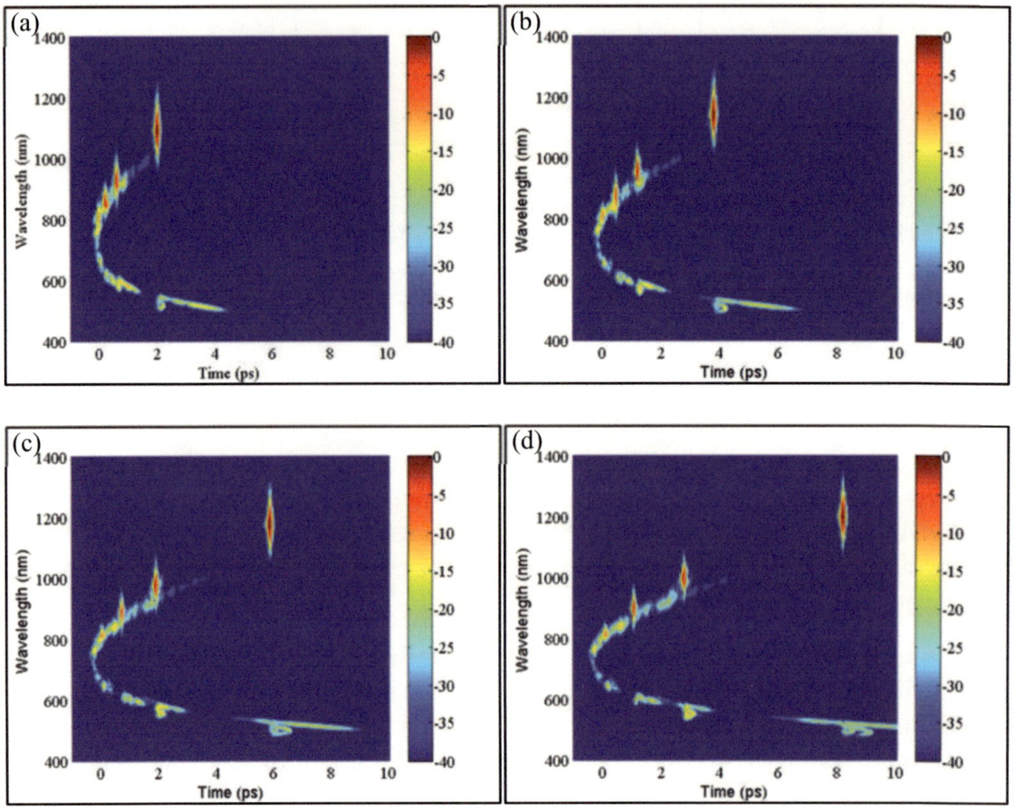

Figure 13. Spectrograms at different propagation distances for input pulse at 800 nm with peak power 10kW and pulse duration 50fs (a) 10 cm; (b) 15 cm; (c) 20 cm; (d) 25 cm;

Although soliton trains will shift towards long-wavelength side due to SSFS [80,81], the pulse duration and spectral bandwidth of first-order soliton are nearly invariable. Figure 14 shows the temporal and spectal profiles of the first-order soliton with different propagating distance with (a) (d) at 10 cm, (b) (e) at 15 cm and (c) (f) at 25 cm. The first-order soliton is selected as both the pump and Stokes just for it has shorter duration and wider spectral bandwidth. It is estimated that the soliton has a pulse duration about 30 fs and spectral bandwidth about 47 nm with 10 cm propagation distance. With further propagation, the pulse duration will increase from 30 to 42 fs, accompanied with the spectral bandwidth decreasing from 47 to 41 nm induced by high-order dispersive and nonlinear effects. Of course the bandwidth is increasing with improving the peak power, but quite few.

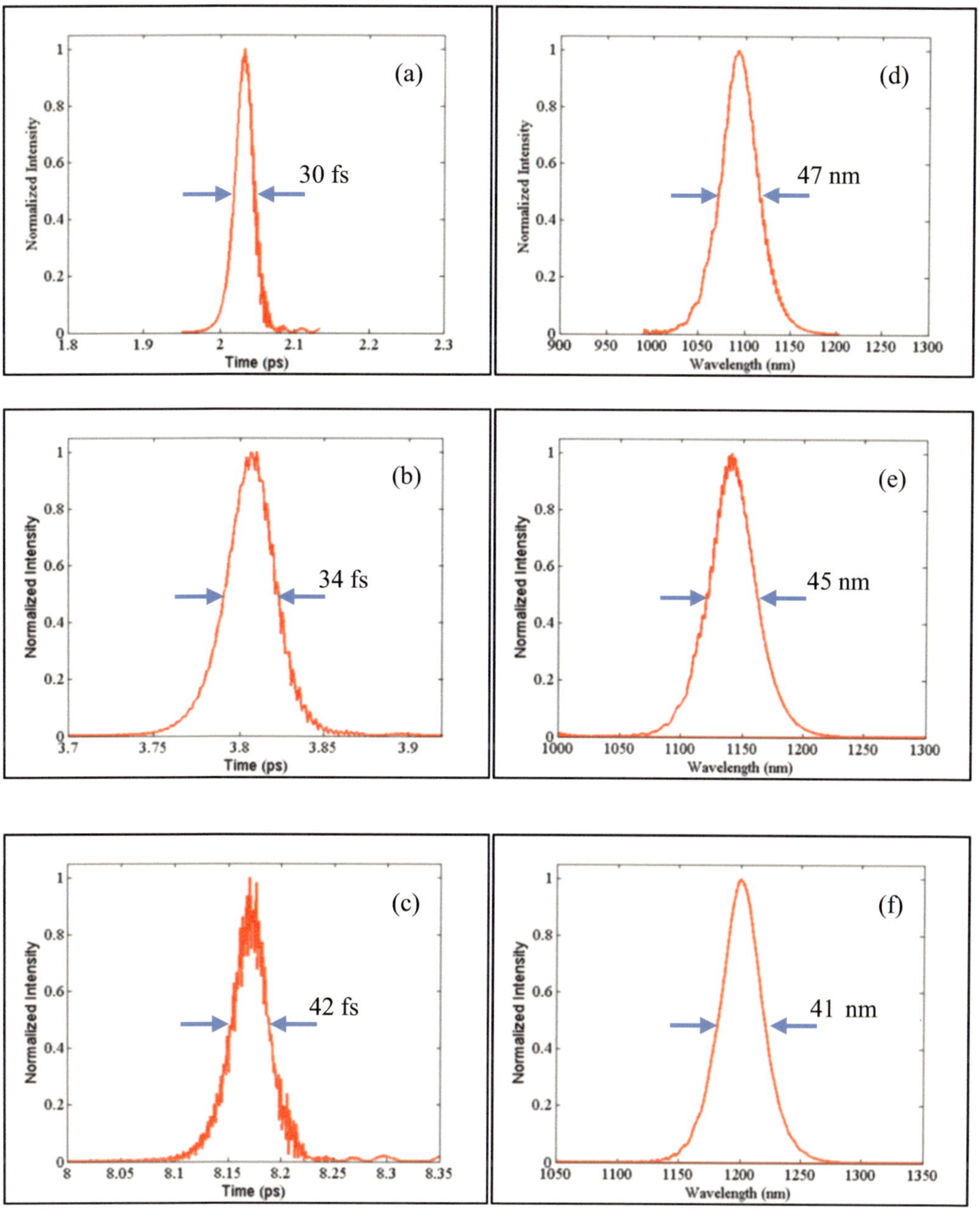

Figure 14. (a ~ c) temporal and (d ~ f) spectral structure of the first-order soliton with (a) (d) at 10 cm, (b) (e) at 15 cm and (c) (f) at 25 cm

Though this first-order soliton shows good temporal and spectral structures, the limited spectral bandwidth makes it not the best choice for CARS microspectroscopy. That is because the pump and Stokes (soliton pulse) can resonant only few molecular bonds instead of the whole bonds, leading to a CARS spectrum with few Raman distinct peaks. To acquire all the Raman distinct peaks simultaneously, SC with both good temporal structure and broad spectral bandwidth is needed.

3.3.2. SC with all-normal dispersion PCF

Recent studies show that SC with broad spectral bandwidth, uniform intensity profile and good coherence is obtained by launching ultrashort laser pulses into an all-normal-dispersion PCF [82,83]. These PCFs exhibit a convex dispersive profile which is flat near the maximum dispersion wavelength (MDW). And they possess none ZDW point in the visible and near-infrared spectral region, known as all-normal dispersion. Without anomalous dispersion region, all the soliton-related dynamics are suppressed, including breakup of input pulse into several pulse trains. And the generated SC is not so sensitive to fluctuations of input pulse and shot noise of pump laser [84]. Because SC generation in these PCFs is dominated by the SPM effect and optical wave breaking, the SC pulse can keep characteristic of a single pulse during the propagation. And the SC pulse is easy for dispersion compensation with such smooth phase distribution. These properties make the generated SC in all-normal dispersion PCFs especially suit for time-resolved applications, such as CARS spectroscopy.

In the following simulations, an all-normal dispersion PCF (from NKT Photonics) with hole diameter 0.5616 μm and hole pitch 1.44 μm is used. Figure 15 shows the cross-section and dispersion profile of the all-normal dispersion PCF respectively. From figure (b), it is clear that this fiber shows all-normal dispersion in the range from visible to near-infrared. And the dispersion parameter D is estimated about -10 ps/nm • km at wavelength about 1060 nm, at which point the nonlinear coefficient is about 37 $W^{-1}m^{-1}$.

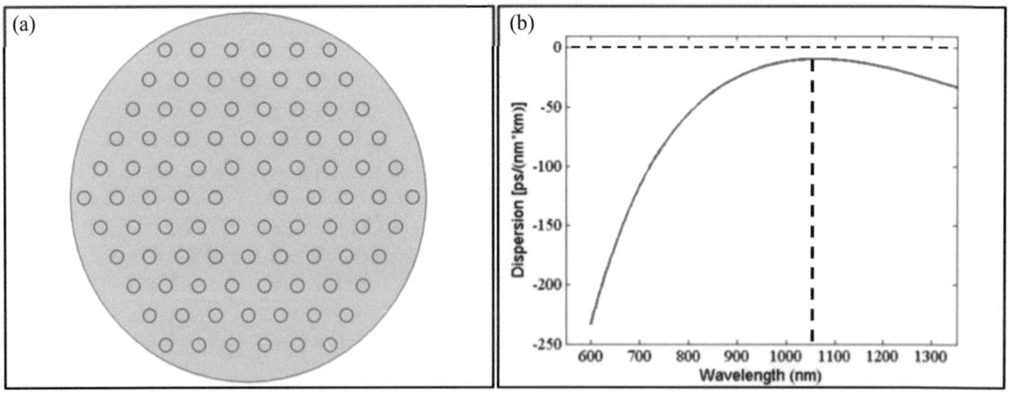

Figure 15. (a) Cross-section model used in simulation and (b) dispersion profile of the all-normal dispersion PCF

Temporal and spectral evolutions with propagating distance are shown in figure 16 when hyperbolic secant pulses centered at 1060 nm with pulse duration 100 fs and peak power 30 kw are injected into the PCF. From figure 16 (b), the spectral broadening is first concluded at the short wavelength through propagation about 38 mm, then at the long wavelength side about 56mm. The spectrum will gain no more broadening with further propagation, except that the generated SC has a smoother spectral profile. In temporal domain, the pulse duration changes a little in the initial stage, but increases gradually with further propagation. And the input pulse keeps its character of single pulse all the time in propagation.

To focus more detailed structures in temporal and spectral domain, evolution slices at different propagating distances are plotted in figure 17 with linear scales. The most notable feature in time domain lies in the conservation of a single pulse during the whole propagation. Although the pulse is broadening during the propagation, self-steepening effect leads to a steeper trailing edge which is clear in slices within 3 centimeters' propagation. After that, the pulse edges becomes steeper which makes the pulse shape like a rectangular function, because even weak dispersion effect also brings about significant changes of pulse shape under abundant SPM-induced frequency chirp. Oscillating structure near the trailing edge as shown in the left graph with 3cm propagation derives from pulse distortion induced by high-order dispersion. And the oscillations near pulse edges are related to optical wave breaking [85].

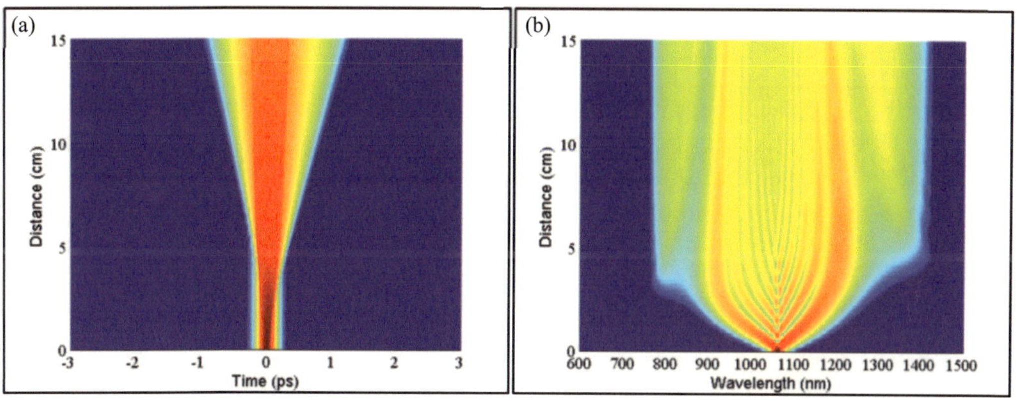

Figure 16. Temporal (left) and spectral (right) evolutions over propagation distance for input pulse centered at 1060 nm with peak power 30 kW and pulse duration 100 fs

In spectral domain, the spectral broadening induced by SPM effect is significant with a clear oscillating structure through the full wavelength. The SPM-induced spectral broadening spreads out towards two sides of the pump wavelength with red-shift components near the leading edge and blue-shift components near the trailing edge. And in the later propagation, the oscillating structure is explained by optical wave breaking. The red-shift components near the leading edge transfer faster and catch up with the red components near leading edge, leading to the interference between them. The blue-shift components near the trailing edge are just an opposite to that. This oscillating structure in spectral domain and the oscillations near pulse edge in the temporal domain are manifestations of the same phenomenon.

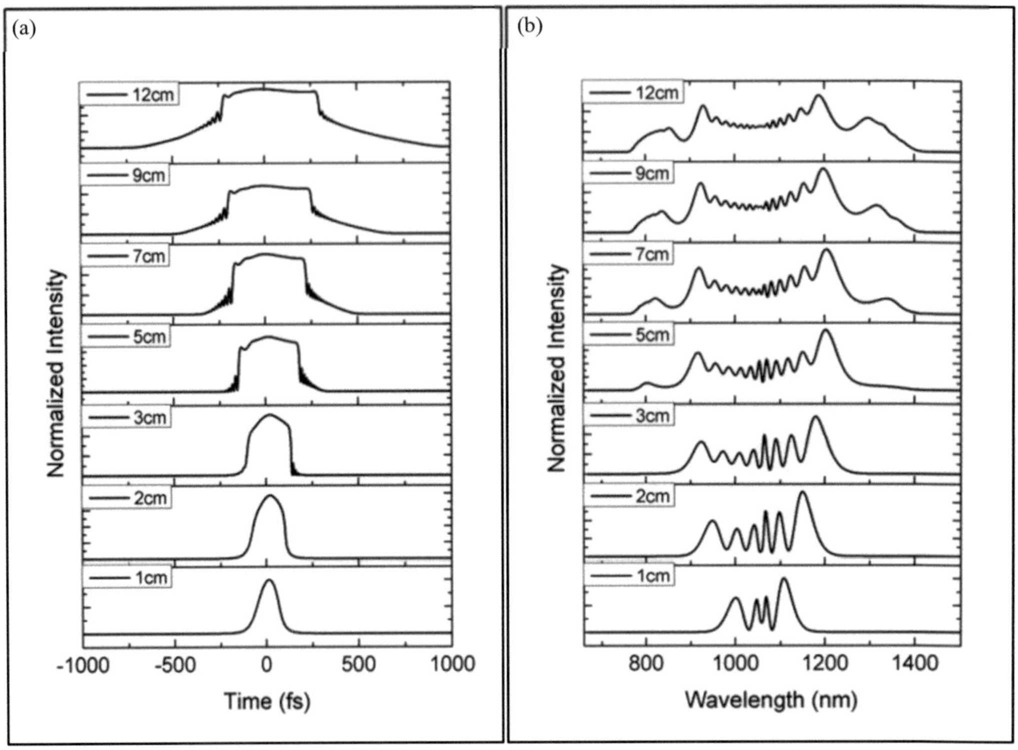

Figure 17. Evolution slices at different propagation distances in temporal (left) and spectral (right) domain

Although SC generated in all-normal dispersion PCF has good temporal and spectral structures, it is useful to study the impact of input pulses on the SC generation, such as pulse duration, pulse energy and so on.

First, the central wavelength, pulse duration and fiber length are restricted to show the impact of pulse energy on SC generating. The input pulses are centered at 1060 nm with pulse duration 100 fs and propagation distance 15 cm. Figure 18(a) shows the variations of spectral bandwidth with different pulse energies ranging from 2 nJ to 10 nJ. It is clear that the generated SC shows larger spectral broadening, much more than 600 nm with pulse energy 4 nJ. For a higher pulse energy, the SC generating is much influenced by the SPM and self-steepening effects, leading to the distinct peaks on each side of the pump wavelength. But the intensity near the pump dips more due to them. The spectral bandwidth is increasing, but the increment becomes small, with pulse energy.

Figure 18(b) shows the impact of pulse duration on generated SC with peak power being a constant 30 kW. The relation between peak power and pulse duration is $E=P \bullet \Delta\tau$ with E being pulse energy, P and $\Delta\tau$ being peak power and pulse duration at half maximum. It is clear that the spectral broadening changes few with different pulse durations. The spectrum reaches 600 nm when the peak power keeps 30 kW. With larger pulse duration, the oscillating structures

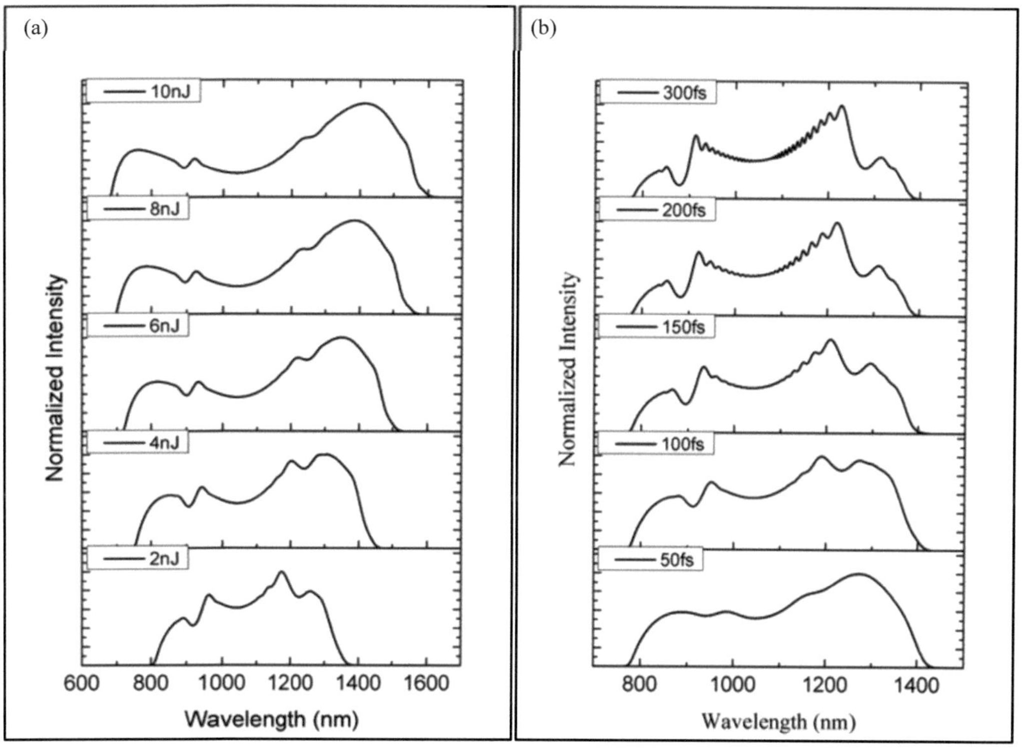

Figure 18. Spectral broadening with different pulse (a) energies and (b) durations

in output spectrum are more obvious. Also the dips near pump wavelength are much more clear. Due to optical wave breaking induced four-wave mixing, the sidebands at each short and long wavelength side are significant, and their peaks are moving towards short and long wavelength direction respectively.

Although the generated SC has a larger spectral broadening when the PCF is pumped near the MDW, the impacts of central wavelength of input pulses on SC generation are analyzed. In figure 19, temporal and spectral structures of generated SC with different pump wavelengths are plotted where the pulse duration, pulse energy and propagation distance are fixed to 100 fs, 2 nJ and 15 cm respectively. For 1060 nm pump, the spectrum is approximately symmetric from 800 nm to 1350 nm with clear oscillating structures near the pump wavelength. For pump wavelength far from the MDW, not only the full spectrum is shorter that that with 1060 nm pump, but the spectrum shows a tail towards the MDW with much lower amplitude than the peaks induced by SPM. There are obviously small tails near the leading edge for 1200 nm pump and trailing edge for 900 nm pump in figure (a) which are induced by high-order dispersion. These tails will apply additional difficulty to dispersion compensation.

From above analyses and discussions, the all-normal dispersion PCF should be pump near the maximum-dispersion point with proper peak power and pulse duration to generate an excellent SC suitable for CARS spectroscopy with broad spectral bandwidth, uniform spectral

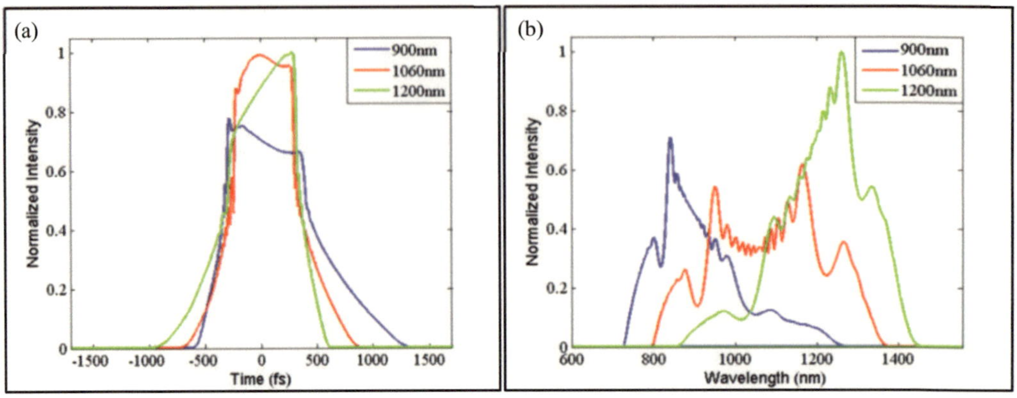

Figure 19. Pulse shapes with different pump wavelengths in (a) temporal and (b) spectral domain

intensities and smooth and excellent temporal structure with smooth phase distribution. The spectral intensity profile is much more smooth with shorter pulse duration under the same peak power. Although higher peak power is beneficial to spectral broadening, it often causes strong tails near the pulse edges and bad flatness of spectral intensity profile. And the fiber length is also very important, because a longer fiber length makes a more flat spectral profile while temporal structure becomes bad.

Based on the basic experimental facilities, pulses centered at 1060 nm with pulse duration 50 fs and pulse energy 2nJ are injected into the all-normal dispersion PCF. The spectrum of output pulse can be expanded to 600~ nm (from ~800 nm to 1400~ nm) after 12 cm propagation with the spectrogram of generated SC being displayed in figure 20(a). This SC shows good spectral continuity and uniformity with tails on both the long and short wavelength sides caused by high-order dispersion. Figure 20(b) demonstrates the temporal distribution of the generated SC pulse with slightly asymmetric structure towards the tailing edge. And the tails on long and short wavelength sides are related to the pedestals near leading and trailing edges. From figure (b), the delay time between different components is less than 2 ps and the temporal phase is predominantly quadratic, which makes the dispersion compensation much easier.

To exploit this continuum pulse into CARS spectroscopy, it is necessary to compensate the delay time induced by dispersive effect and other nonlinear effects, leading to a same propagating speed of different frequency components. In terms of such small dispersion, a pair of prisms made of BK7 glass are used to compensate it. The spectrogram and temporal, spectral profiles after dispersion compensation are plotted in figure 21, with (a) being the spectrogram, (b) the temporal profile and (c) the spectral profile. After dispersion compensation, most components in the generated SC are propagating at a same speed, except the tails on long and short wavelength sides, for the prisms are not effective enough to compensation of high-order dispersion. From figure (b), the pulse duration of generated SC is compressed to about 15 fs from initial ~2 ps in spite of a little oscillation near the pulse edges. Figure (c) shows the spectral profile on a linear scale with bandwidth over 600 nm. And the spectral intensity possesses a good uniformity.

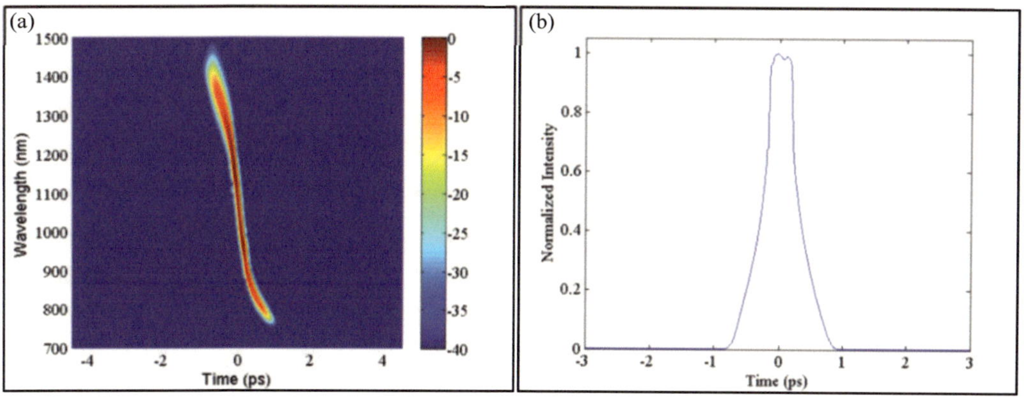

Figure 20. (a) Spectrogram and (b) temporal profile of the generated SC pulse

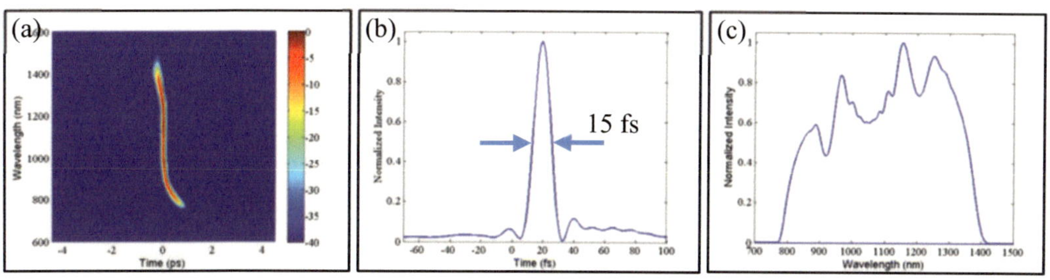

Figure 21. (a) Spectrogram, (b) temporal profile and (c) spectral profile after dispersion compensation

Supposing that the delay time between different frequency components is limited to less than 40 fs, it is estimated that the useful components still span over 400 nm from 900 to 1300 nm. This SC after dispersion compensating shows wonderful temporal structures, broad spectral bandwidth, and uniform spectral intensity, which make it well suited for time-resolved applications, especially CARS microspectroscopy. To make the best of the tails in figure (a), spatial light modulator is suggested for the high-order dispersion compensation [86,87].

4. Supercontinuum used in infrared nanoscopy

4.1. Theory of Infrared absorption spectroscopy

The infrared (IR) absorption spectroscopy is an excellent method for biological analyses. It enables the nondestructive, label-free extraction of biochemical information and images [27,88]. The physical basis of IR absorption is that, when an infrared light interacts with the molecule, the vibrations or rotations in a molecule may cause a net change in the dipole moment. The alternating electrical field of the radiation interacts with fluctuations in the dipole moment of the molecule. If the frequency of the radiation matches the vibrational frequency of the molecule, then the radiation will be absorbed [89,90]. Every material has its unique IR absorption spectra that can be used as a "fingerprint" for identification. Absorption in the infrared region results in changes in vibrational and rotational status of the molecules. The absorption frequency depends on the vibrational frequency of the molecule, whereas the absorption intensity depends on how effectively the infrared photon energy can be transferred to the molecule, and on the change in the dipole moment that occurs as a result of molecular vibration [91-93]. In a word, a molecule will absorb infrared light only if the absorption causes a change in the dipole moment. Infrared spectroscopy is widely applied to various samples such as liquid, gas, and solid-state matter to identify and to quantify the unknown materials. It is an effective technique to identify compounds and is used extensively to detect functional groups.

Collecting of IR spectra signal usually occurs in two different ways. One is the conventional dispersive-type spectrometer, which employs a grating or a prism to disperse light into individual frequencies, and a slit placed in front of the detector to determine which frequency to reach the detector [94]. However, dispersive-type spectrometer is outdated today. Modern IR spectrometers are the so-called Fourier transform infrared absorption (FTIR) spectroscopy instruments [95,96]. All the FTIR spectrometer operates on the Fourier transform principle, and refers to the manner in which the data is collected and converted from an interference pattern to a spectrum. FTIR spectrometers have progressively replaced the dispersive instruments for most applications due to their superior speed and sensitivity. They have greatly extended the capabilities of IR spectroscopy and have been applied to many fields that are very difficult or nearly impossible to study by dispersive instruments [97,98]. FTIR can be utilized to measure some components of an unknown mixture and is currently applied to the analysis of solids, liquids, and gases.

The Basic FTIR System as shown in figure 22, there are mainly four parts: Light source, interferometer, sample compartment, and the detector. The beam splitter divides the incoming infrared beam into two beams. One beam reflects off the fixed mirror, and the other beam reflects off the moving mirror, constituting a Michelson interferometer. After reflected from the two mirrors, the divided beams meet each other again at the beam splitter. Approximately 50% of the light passes through the beam splitter and is reflected back along its path by a fixed mirror, and half of the light is reflected by the same beam splitter. The other 50% fraction of the incident light is reflected onto a moving mirror. Light from the moving mirror returns along its original path and half of the light intensity is transmitted through the beam splitter. The two beams meet again at the same beam splitter and then pass through the sample cell, to the infrared sensor. Thus 25% of the incident light from the source reaches the sensor from the fixed mirror and 25% from the moving mirror. As the path length of the two light beams striking the sensor will differ, there will be destructive and constructive interference.

The beam from the moving mirror has traveled a different distance than the beam from the fixed mirror. When the two beams are combined by the beam splitter, an interference pattern is created, since some of the wavelengths recombine constructively and some destructively. This interference pattern is called an interferogram. This interferogram then goes from the

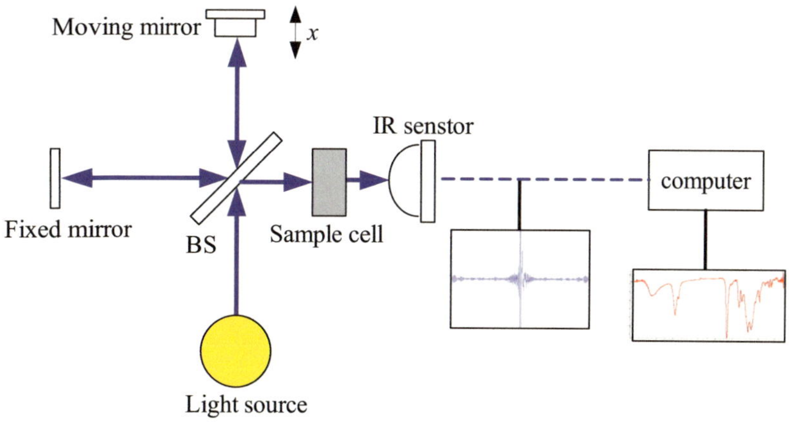

Figure 22. The main components of a FTIR

beam splitter to the sample, where some energy is absorbed and some is transmitted. The transmitted portion reaches the detector. Every stroke of the moving mirror in the interferometer equals one scan of the entire IR source spectrum, and individual scan can be combined to give better representation of the actual absorbance of the sample. Each point in the interferogram contains information from each wavelength of light being measured. The detector reads information about every wavelength in range of the IR source simultaneously. To obtain the infrared spectrum, the detector signal is sent to the computer, and an algorithm called a Fourier transform is performed on the interferogram to convert it into a spectrum. This transmittance spectrum can be converted to absorbance. The process of collecting the IR spectrum in an FTIR spectrometer is illustrated in figure 23. FTIR spectrometer simultaneously collects spectral data in a wide spectral range of the IR source [99]. In contract, in a dispersive spectrometer, every wavelength across the spectrum must be measured individually. This is a slow process, and typically only one measurement scan of the sample is made. Accordingly, the superior speed and sensitivity of FTIR is the significant advantage over a conventional dispersive spectrometer which measures intensity over a narrow range of wavelengths at a time [100,101].

The FTIR refers to the manner in which the data is collected and converted from an interference pattern to a spectrum. The FTIR spectrometer operates on a principle called Fourier transform. The mathematical expression of Fourier transform can be expressed as

$$F(\omega)=\int_{-\infty}^{+\infty} f(x)e^{i\omega x}dx \tag{38}$$

And the reverse Fourier transform is

$$f(x)=\frac{1}{2\pi}\int_{-\infty}^{+\infty} f(x)e^{-i\omega x}d\omega \tag{39}$$

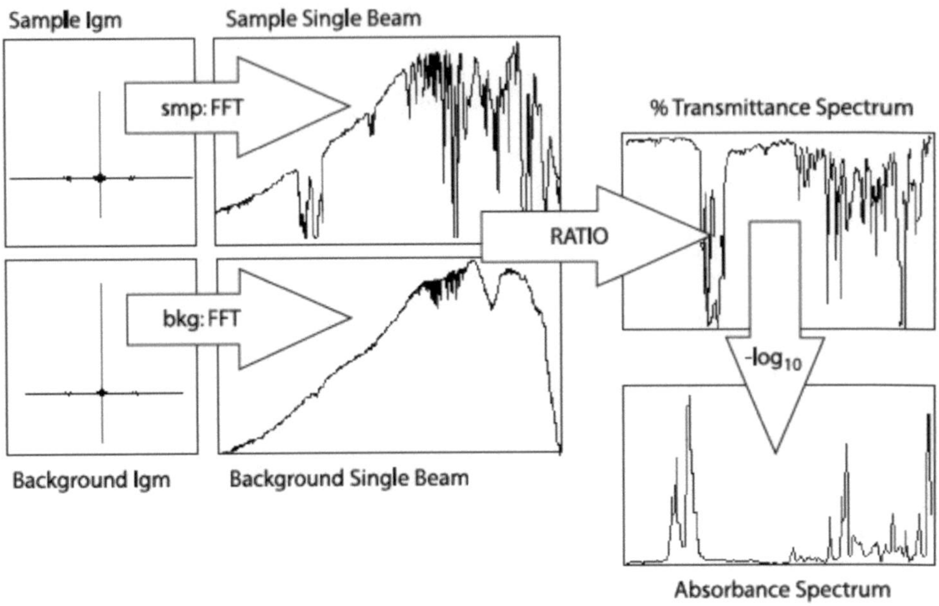

Figure 23. The process of collecting the IR spectrum in an FTIR spectrometer

where ω is angular frequency and x is the optical path difference. $F(\omega)$ is the spectrum and $f(x)$ is called the interferogram. It is clear that if the interferogram $f(x)$, is determined experimentally by using the Michelson interferometer, the spectrum $F(\omega)$ can be obtained by using Fourier transform. The detector receives a signal, i.e., the interferogram, which is a summation of all the interferences resulting from the constructive and destructive interaction between each wavelength component and all the others. The two most popular detectors for a FTIR spectrometer are deuterated triglycine sulfate (pyroelectric detector) and mercury cadmium telluride (photon or quantum detector). The measured signal is sent to the computer where the Fourier transformation takes place.

FTIR spectrometer simultaneously collects spectral data of the IR in a wide spectral range. This confers a significant advantage over a dispersive spectrometer which measures intensity over a narrow range of wavelengths at a time. FTIR spectrometers have progressively replaced dispersive instruments for most applications due to their superior speed and sensitivity, and opens up new applications of infrared spectroscopy

4.2. SC used in IR nanoscopy

FTIR spectroscopy is a widely used analytical tool for chemical identification of inorganic, organic, and biomedical materials, as well as for exploring conduction phenomena [102-104]. Because of the diffraction limit, however, conventional FTIR cannot achieve nano-scaled resolution [105]. Therefore a FTIR system that allows for infrared-spectroscopic nano-imaging is required. The ability to use a spectrally broad source in a FTIR system makes this technique promising at the nano-scale for the chemical identification of unknown nanostructures. Relying on the use of scattering-type scanning near-field optical microscopy [106,107] (s-SNOM), the diffraction limit of a FTIR system is overcome by the use of a sharp atomic force microscope (AFM) tip acting as an antenna to concentrate the incident radiation to nano-scaled volumes [108-110]. Recording of the tip-scattered light thus yields nano-scaled resolved optical images. With the introduction of s-SNOM, the diffraction-limited resolution in FTIR microscopy can be overcome by several orders of magnitude. Nano-FTIR has interesting application potential in widely different sciences and technologies.

Traditional FTIR spectrometers are based on either thermal IR sources with low brightness and coherence, such as a globar or Hg-lamp, or synchrotron radiation. The synchrotron radiation is complex, expensive, power consuming and always accompanied with intensity fluctuation in spite of high brightness and coherence [29,111]. The advent of non-silica fibers brings us a new IR source with both high brightness and coherence, not to mention the broad bandwidth [31,33,112,113]. Besides fused silica, the PCFs can also be made of other materials, such as chalcogenide and fluoride which show novel characteristics near mid-infrared, such as high transmissivity. Based on these materials, the PCF can shift its ZDW to mid-infrared region, which makes the SC generation in mid-infrared realized easily [114]. And their intrinsic high IR transmissivity and nonlinear coefficient makes it especially suit for mid-infrared SC generation. With these materials, step-index fibers, tapered fibers and photonic crystal fiber can be used to generate the mid-infrared SC [115,116]. Under properly pumping condition, the generated SC shows an ultra-broad spectrum range from 2 μm to 12 μm. These mid-infrared supercontinuum possess a broad spectral bandwidth, higher brightness and coherence [117]. The nano-FTIR combing with such a mid-infrared supercontinuum source, has interesting application potential in widely different sciences and technologies, ranging from the semiconductor industry to nanogeochemistry.

In the following paragraphs, we will discuss the process of SC generation with the widely used PCF made of chalcogenide glass. It is necessary to figure out the refraction index and nonlinear response, because they often vary a lot for different materials. The refraction index of bulk As_2Se_3 glass is approximate to the Sellmeier equation as in Eq. (21). The approximate profile show good agreement with measured data [118].

$$n^2(\lambda)=1+\lambda^2\left[\frac{A_0^2}{\lambda^2-A_1^2}+\frac{A_2^2}{\lambda^2-19^2}+\frac{A_4^2}{\lambda^2-4A_1^2}\right] \tag{40}$$

with $n(\lambda)$ being the refraction index, λ the input wavelength expressed in microns. The related Sellmeier coefficients are A_0=2.234921, A_1=0.24164, A_2=0.347441, A_4=1.308575. By calculation, the variations of refractive index and dispersion parameter with wavelength are shown in figure 24(a) with n being the refractive index, n_g the group index. It is clear the group index n_g decreases with wavelength below about 7.2 μm and increases beyond that wavelength point. Figure 24(b) exhibits the variation of dispersion with wavelength, where single ZDW point locates in the region from 2 to 14 μm.

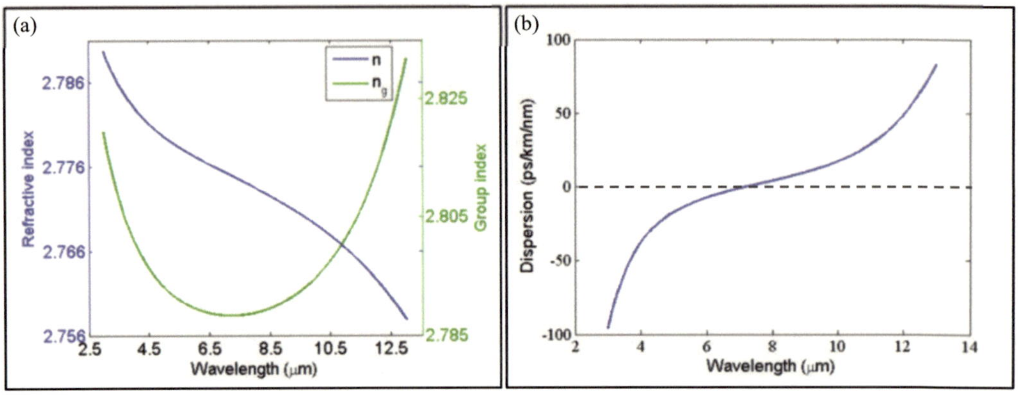

Figure 24. (a) Variation of refractive index n and group index $n_g$with wavelength (b) variation of dispersion parameter with wavelength

Considering the As_2Se_3-based PCF has an air hole diameter about 1.2 μm and a hole pitch about 3.0 μm. The ideal structure used in FEM is plotted as in figure 25(a) with a solid core surrounded by four hexagonal rings of air holes. And figure 25(b) shows the dispersion profile of this PCF with single ZDW about 3.72 μm. To obtain a larger spectral broadening, this PCF is pump in the anomalous dispersion regime near ZDW, saying 3.75 μm at which the nonlinear coefficient is about 2.7 $m^{-1}W^{-1}$. The nonlinear coefficient in As_2Se_3-based PCF is tens to hundreds times larger than that in fused silica due to the higher nonlinear refractive index n_2 about $2.4\times10^{-17}m^2/W$.

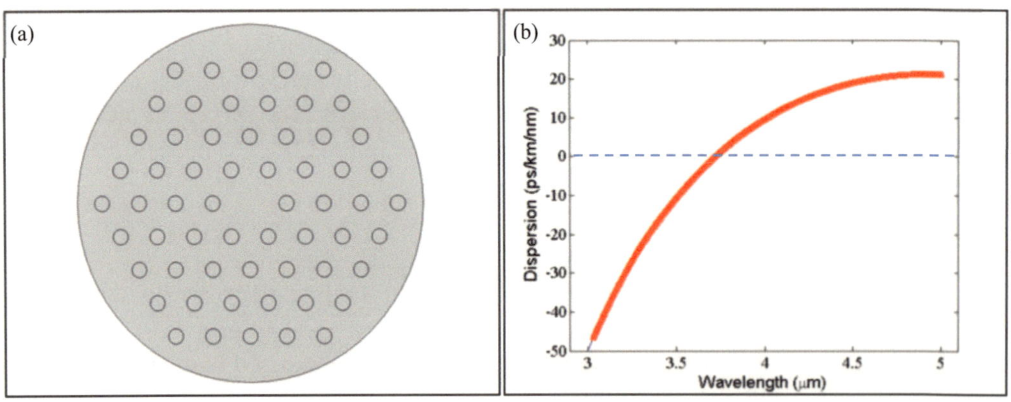

Figure 25. (a) Cross-section structure in simulation and (b) dispersion profile of the PCF

Before modeling the process of SC generation in this As_2Se_3-based PCF, it is necessary to figure out the nonlinear response function $R(t)=(1 - f_R)\delta(t) + f_R h_R(t)$, due to the different materials. For chalcogenide glasses, the fractional contribution of the Raman response to the total nonlinear response is f_R=0.115 and the delayed Raman response function $h_R(t)$ can be expressed as [118,119]

$$h_R(t) = \frac{\tau_1^2 + \tau_2^2}{\tau_1 \tau_1^2} \exp\left(-\frac{t}{\tau_2}\right) sin\left(\frac{t}{\tau_1}\right) \tag{41}$$

with τ_1=23.1 fs, τ_2=195 fs.

In the simulation, the injected pulse has a hyperbolic secant profile with with 200 fs pulse duration and 30 W peak power. The temporal and spectral evolutions of input pulse with propagating distance are plotted in figure 26. It is clear that the output spectral bandwidth reaches 1.6μm from 3 μm to about 4.7μm after propagating about 20 cm. Further propagation brings in only little spectral broadening on the long-wavelength side, which is induced by intrapulse Raman scattering. In time domain, strongly temporal compression occurs within about 11 cm. After that, there is an obvious pulse broadening and soliton fission. Because different frequency components often possess different velocities, time delay between different frequencies becomes larger with a longer propagation,.

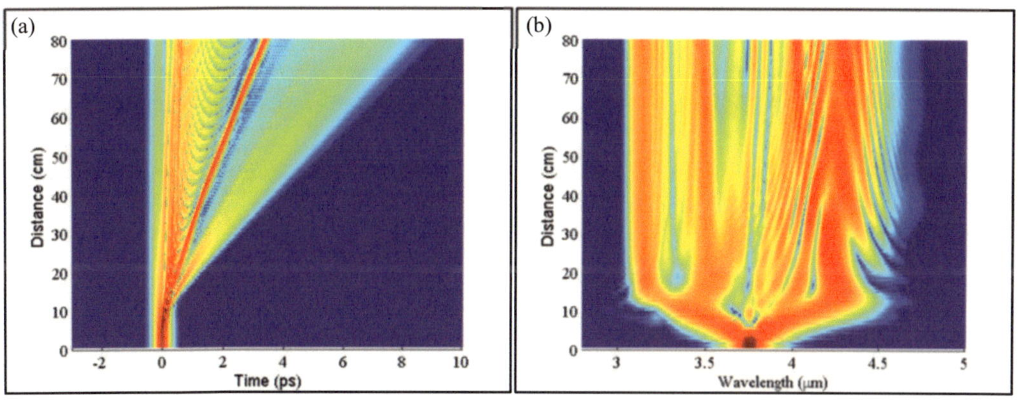

Figure 26. Temporal (a) and spectral (b) evolutions over propagation distance for input pulse centered at 3.75 μm with peak power 30 W and pulse duration 200 fs

To elaborate on the dynamics of SC generation, different spectral slices are exhibited in figure 27(a) on a logarithmic scale. All the spectrums in figure 27(a) are cut off at -80 dB. In the initial propagation stage, spectral broadening is approximately symmetrical. And the main mechanism of spectral broadening is SPM effect. The SPM-induced spectral broadening has a notable oscillation structure with the outermost peaks possessing the most intense as shown in figure. After around 12 cm, the spectral broadening becomes asymmetric with a larger broadening on the long wavelength side. The out side peaks on long- and short-wavelength side are related to soliton fission and dispersive wave generation. With further propagation, spectrum on long-wavelength side spans little due to weak Raman scattering.

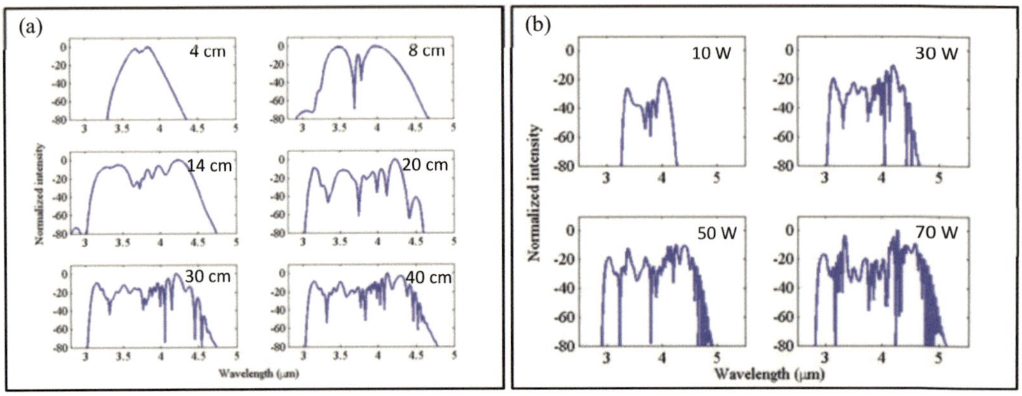

Figure 27. Spectral evolutions for (a) selected propagating distance and (b) peak power

Although the generated mid-SC has a bandwidth of 1.6 μm, it is not enough for drive all the vibraions simultaneously. A feasible approach is to increase the peak power of pump pulse or shorten the pulse duration. Figure 27(b) shows the numerical spectrums for different peak powers after 30 cm propagation with the same pulse duration 200 fs. The input peak powers are 10 W, 30 W, 50 W and 70 W for the top-left, top–right, bottom-left and bottom–right figures respectively. It is clear that the total spectral bandwidth of generated SC is increasing with peak power. And with higher peak power, shorter propagation distance is need to make the spectral broadening to its maximum.

To study the impact of pulse duration on SC generation, peak power of input pulses is kept an constant 30 W in the following simulations. Figure 28(b) exhibit the spectral evolutions for different pulse durations with figures on top-left, top-right, bottom-left, and bottom-right being 100 fs, 200 fs, 300 fs and 500 fs respectively. For pump pulse with shorter duration, the distinct soliton structure is much more clear as in figure (a) and (b) because of a shorter dispersive length L_D. They also have a smaller spectral broadening compared with that in figure (c) and (d) due to a lower pulse energy. However, larger input pulse is associated with longer dispersive length, larger pulse energy and higher soliton order which result in a less clear soliton structure and more complex spectral profile in spite of a larger spectral broadening. In the initial propagation, spectral broadening is more with shorter pulse duration because of the larger SPM-induced chirp and its role in FWM.

The spectral broadening with As_2Se_3-based PCF is much more than that in silica-glass PCF because of the high nonlinearity. To obtain a continuum with both broad spectral bandwidth and good time structure, a shorter propagation distance is suggested because the weak SSFS effect shifts soliton little towards long-wavelength side with finite propagating distance. And with novel fiber structures, the generated SC has a much larger spectral broadening, such as the mid-infrared SC from 2 to 10 μm by Baili *et al* [118].

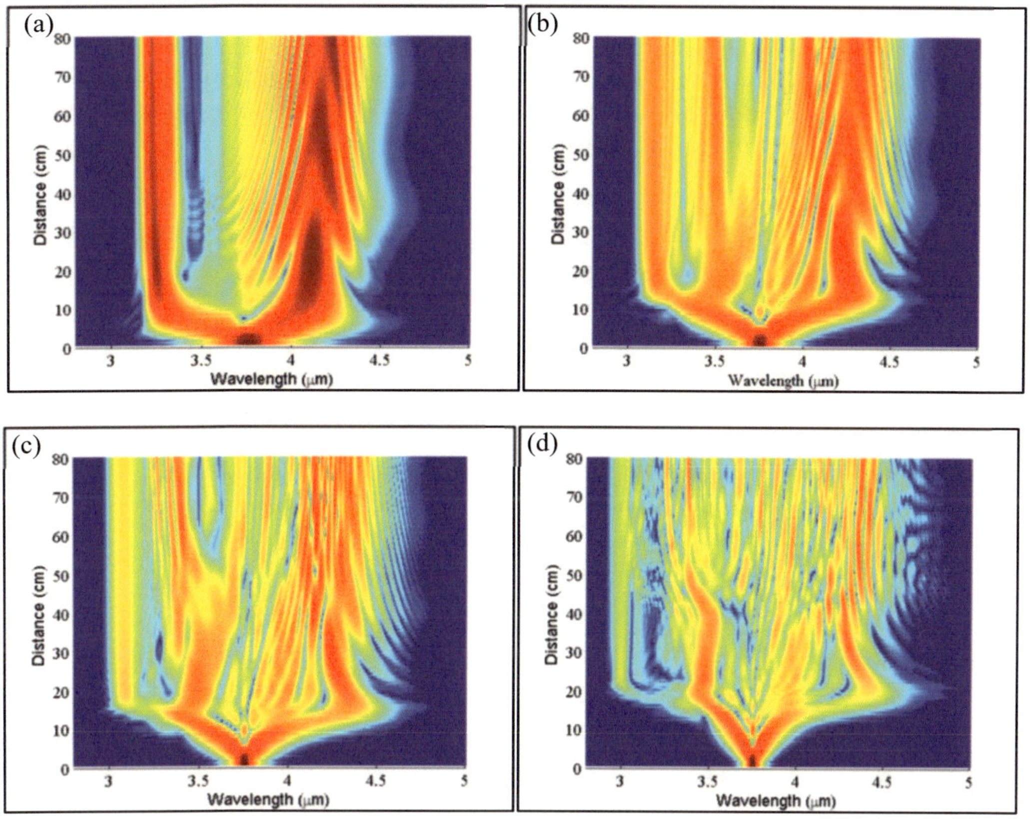

Figure 28. Evolutions over propagation distance for different pulse durations (a) 100 fs; (b) 200 fs; (c) 300 fs; (d) 500 fs

5. Conclusions and prospects

In this chapter, we mainly described the generalized nonlinear Schrödinger equation which is widely used in simulating the pulse evolution in photonic crystal fibers. In this equation, both the higher-order dispersion and nonlinear effects are included, such as self-steepening, optical shock formation and intrapulse Raman scattering. Then predictor-corrector split-step Fourier method are introduced to solve the nonlinear Schrödinger equation. Base the equation and numerical method, process of SC generation in a traditional PCF is analyzed in details. Spectral broadening in anomalous dispersions regime is dominated by soliton dynamics, including soliton fission and soliton self-frequency shift, while it is related to SPM effect and optical wave breaking in normal dispersion region. The generated SC has many applications, such as optical frequency metrology, tunable ultrafast fiber laser and nonlinear microscopy for its broad spectrum, high coherence and uniform intensity. Based on the vibrational spectroscopy, CARS microscopy and IR microscopy has been widely used in cellular biology for their label-free and nondestructive imaging. Although the spatial resolution of CARS microscopy is not such high due to the diffraction limit, the APIPD opens up a quite feasible approach to realize nano-scaled imaging. Furthermore, the combination of this CARS nanoscopy and the supercontinuum generated in an all-normal dispersion PCF is of crucial importance to study the fine structures and metabolic dynamics in live cells. Besides, combining with atomic force microscope, FTIR microscopy has a spatial resolution below 100nm. This nano-FTIR combing with a mid-infrared supercontinuum source generated from the PCF becoming a powerful tool for chemical identification of unknown nanostructures.

Author details

Shuanglong Liu, Wei Liu and Hanben Niu*

*Address all correspondence to: hbniu@szu.edu.cn

Key Laboratory of Optoelectronic Devices and Systems of Ministry of Education and Guangdong Province, College of Optoelectronic Engineering, Shenzhen University, Shenzhen, P.R.China

References

[1] Alfano, R. R.Shapiro, S. L., Emission in the Region 4000 to 7000 Å Via Four-Photon Coupling in Glass, Physical Review Letters, 1970, 24(11): 584-587

[2] Kaiser, P.Astle, H. W., Low-Loss Single-Material Fibers Made From Pure Fused Silica, Bell System Technical Journal, 1974, 53(6): 1021-1039

[3] Knight, J. C., Birks, T. A., Russell, P. S. J. & Atkin, D. M., All-silica single-mode optical fiber with photonic crystal cladding, Opt. Lett., 1996, 21(19): 1547-1549

[4] Russell, P., Photonic crystal fibers, science, 2003, 299(5605): 358-362

[5] Mogilevtsev, D., Birks, T. A. & Russell, P. S. J., Group-velocity dispersion in photonic crystal fibers, Opt. Lett., 1998, 23(21): 1662-1664

[6] Ranka, J. K., Windeler, R. S. & Stentz, A. J., Visible continuum generation in air-silica microstructure optical fibers with anomalous dispersion at 800 nm, Opt. Lett., 2000, 25(1): 25-27

[7] Wadsworth, W. J., Knight, J. C., Ortigosa-Blanch, A., Arriaga, J. *et al.* Soliton effects in photonic crystal fibres at 850 nm. *Electronics Letters* 36, 53-55 (2000).

[8] Liu, X., Xu, C., Knox, W. H., Chandalia, J. K. *et al.*, Soliton self-frequency shift in a short tapered air-silica microstructure fiber, Opt. Lett., 2001, 26(6): 358-360

[9] Husakou, A. V.Herrmann, J., Supercontinuum Generation of Higher-Order Solitons by Fission in Photonic Crystal Fibers, Physical Review Letters, 2001, 87(20): 203901

[10] Washburn, B. R., Ralph, S. E., Lacourt, P. A., Dudley, J. M. *et al.*, Tunable near-infrared femtosecond soliton generation in photonic crystal fibres, ELECTRONICS LETTERS, 2001, 37(25): 1510-1512

[11] Genty, G., Lehtonen, M., Ludvigsen, H., Broeng, J. *et al.*, Spectral broadening of femtosecond pulses into continuum radiation in microstructured fibers, Opt. Express, 2002, 10(20): 1083-1098

[12] Avdokhin, A. V., Popov, S. V. & Taylor, J. R., Continuous-wave, high-power, Raman continuum generation in holey fibers, Opt. Lett., 2003, 28(15): 1353-1355

[13] Dudley, J. M., Genty, G. & Coen, S., Supercontinuum generation in photonic crystal fiber, Reviews of Modern Physics, 2006, 78(4): 1135-1184

[14] Wisse, E., De Zanger, R., Jacobs, R. & McCuskey, R., Scanning electron microscope observations on the structure of portal veins, sinusoids and central veins in rat liver, Scanning electron microscopy, 1982, Pt 3): 1441-1452

[15] Binnig, G.Rohrer, H., Scanning tunneling microscopy, Surface Science, 1983, 126(1): 236-244

[16] Binnig, G., Gerber, C., Stoll, E., Albrecht, T. *et al.*, Atomic resolution with atomic force microscope, EPL (Europhysics Letters), 1987, 3(12): 1281

[17] Juhun, P., Jinwoo, L., Seon, N., Kwang, H. *et al.*, Sub-Diffraction Limit Imaging of Inorganic Nanowire Networks Interfacing Cells, Small, 2014, 10(3): 462-468

[18] Hess Samuel T, G. T. P. K., Mason Michael D, Ultra-High Resolution Imaging by Fluorescence Photoactivation Localization Microscopy, Biophysical Journal, 2006, 91(11): 4258-4272

[19] Michael J Rust, M. B., Xiaowei Zhuang, Sub-diffraction-limit imaging by stochastic optical reconstruction microscopy (STORM), Nature Methods, 2006, 3(10): 47

[20] Hell Stefan W, W. J., Breaking the diffraction resolution limit by stimulated emission: stimulated-emission-depletion fluorescence microscopy, Opt. Lett., 1994, 19(11): 780-782

[21] Wu, J., Cohen, L. & Mason, W., Fluorescent and luminescent probes for biological activity, London: Academic, 1993,

[22] Evans, C. L.Xie, X. S., Coherent Anti-Stokes Raman Scattering Microscopy: Chemical Imaging for Biology and Medicine, Annual Review of Analytical Chemistry, 2008, 1(1): 883-909

[23] Yuan, J.-H., Xiao, F.-R., Wang, G.-Y. & Xu, Z.-Z., Imaging properties of coherent anti-Stokes Raman scattering microscope, Chinese Physics, 2005, 14(935-941

[24] Xia, L., Hui, Z., Xiang-Yun, Z., Shi-An, Z. *et al.*, Experimental Investigation on Selective Excitation of Two-Pulse Coherent Anti-Stokes Raman Scattering, Chinese Physics Letters, 2008, 25(6): 2062

[25] Liu Wei, N. H., Diffraction barrier breakthrough in coherent anti-Stokes Raman scattering microscopy by additional probe-beam-induced phonon depletion, Physical Review A, 2011, 83(2): 023830

[26] Sage, J. T., Zhang, Y., McGeehan, J., Ravelli, R. B. G. *et al.*, Infrared protein crystallography, Biochimica et Biophysica Acta (BBA) - Proteins and Proteomics, 2011, 1814(6): 760-777

[27] Gillie, J. K.Hochlowski, J., Infrared Spectroscopy, Analytical Chemistry, 2000, 75(12):

[28] Huth, F., Schnell, M., Wittborn, J., Ocelic, N. *et al.*, Infrared-spectroscopic nanoimaging with a thermal source, Nature materials, 2011, 10(5): 352-356

[29] Reffner, J. A., Martoglio, P. A. & Williams, G. P., Fourier transform infrared microscopical analysis with synchrotron radiation: the microscope optics and system performance, Review of Scientific Instruments, 1995, 66(2): 1298-1302

[30] Tobin, M. J., Chesters, M. A., Chalmers, J. M., Rutten, F. J. *et al.*, Infrared microscopy of epithelial cancer cells in whole tissues and in tissue culture, using synchrotron radiation, Faraday discussions, 2004, 126(27-39

[31] Kumar, V., George, A., Knight, J. & Russell, P., Tellurite photonic crystal fiber, Opt. Express, 2003, 11(20): 2641-2645

[32] Sanghera, J. S., Shaw, L. B. & Aggarwal, I. D., Applications of chalcogenide glass optical fibers, Comptes Rendus Chimie, 2002, 5(12): 873-883

[33] Xia, C., Kumar, M., Kulkarni, O. P., Islam, M. N. *et al.*, Mid-infrared supercontinuum generation to 4.5 μm in ZBLAN fluoride fibers by nanosecond diode pumping, Opt. Lett., 2006, 31(17): 2553-2555

[34] Diament, P. *Wave transmission and fiber optics*. Vol. 74 (Macmillan, 1990).

[35] Blow, K. J.Wood, D., Theoretical description of transient stimulated Raman scattering in optical fibers, Quantum Electronics, IEEE Journal of, 1989, 25(12): 2665-2673

[36] Karasawa, N., Nakamura, S., Nakagawa, N., Shibata, M. *et al.*, Comparison between theory and experiment of nonlinear propagation for a-few-cycle and ultrabroadband optical pulses in a fused-silica fiber, Quantum Electronics, IEEE Journal of, 2001, 37(3): 398-404

[37] Kibler, B., Dudley, J. M. & Coen, S., Supercontinuum generation and nonlinear pulse propagation in photonic crystal fiber: influence of the frequency-dependent effective mode area, Appl. Phys. B, 2005, 81(2-3): 337-342

[38] Mamyshev, P. V.Chernikov, S. V., Ultrashort-pulse propagation in optical fibers, Opt. Lett., 1990, 15(19): 1076-1078

[39] Chernikov, S. V.Mamyshev, P. V., Femtosecond soliton propagation in fibers with slowly decreasing dispersion, J. Opt. Soc. Am. B, 1991, 8(8): 1633-1641

[40] Stolen, R. H.Tomlinson, W. J., Effect of the Raman part of the nonlinear refractive index on propagation of ultrashort optical pulses in fibers, J. Opt. Soc. Am. B, 1992, 9(4): 565-573

[41] Lin, Q.Agrawal, G. P., Raman response function for silica fibers, Opt. Lett., 2006, 31(21): 3086-3088

[42] Hollenbeck, D.Cantrell, C. D., Multiple-vibrational-mode model for fiber-optic Raman gain spectrum and response function, J. Opt. Soc. Am. B, 2002, 19(12): 2886-2892

[43] Xueming, L.Byoungho, L., A fast method for nonlinear Schrodinger equation, Photonics Technology Letters, IEEE, 2003, 15(11): 1549-1551

[44] Reid, D. T., Loza-Alvarez, P., Brown, C. T. A., Beddard, T. *et al.*, Amplitude and phase measurement of mid-infrared femtosecond pulses by using cross-correlation frequency-resolved optical gating, Opt. Lett., 2000, 25(19): 1478-1480

[45] Linden, S., Kuhl, J. & Giessen, H., Amplitude and phase characterization of weak blue ultrashort pulses by downconversion, Opt. Lett., 1999, 24(8): 569-571

[46] Dudley, J., Gu, X., Xu, L., Kimmel, M. *et al.*, Cross-correlation frequency resolved optical gating analysis of broadband continuum generation in photonic crystal fiber: simulations and experiments, Opt. Express, 2002, 10(21): 1215-1221

[47] Kane, D. J.Trebino, R., Single-shot measurement of the intensity and phase of an arbitrary ultrashort pulse by using frequency-resolved optical gating, Opt. Lett., 1993, 18(10): 823-825

[48] Trebino, R., DeLong, K. W., Fittinghoff, D. N., Sweetser, J. N. *et al.*, Measuring ultrashort laser pulses in the time-frequency domain using frequency-resolved optical gating, Review of Scientific Instruments, 1997, 68(9): 3277-3295

[49] Tatian, B., Fitting refractive-index data with the Sellmeier dispersion formula, Appl. Opt., 1984, 23(24): 4477-4485

[50] Adams, M., An introduction to optical waveguides, Chichester-New York, 1981,

[51] Malitson, I. H., Interspecimen Comparison of the Refractive Index of Fused Silica, J. Opt. Soc. Am., 1965, 55(10): 1205-1208

[52] Cucinotta, A., Selleri, S., Vincetti, L. & Zoboli, M., Holey fiber analysis through the finite-element method, Photonics Technology Letters, IEEE, 2002, 14(11): 1530-1532

[53] Koshiba, M., Full-vector analysis of photonic crystal fibers using the finite element method, IEICE transactions on electronics, 2002, 85(4): 881-888

[54] Brechet, F., Marcou, J., Pagnoux, D. & Roy, P., Complete analysis of the characteristics of propagation into photonic crystal fibers, by the finite element method, Optical Fiber Technology, 2000, 6(2): 181-191

[55] Shibata, N., Braun, R. & Waarts, R., Phase-mismatch dependence of efficiency of wave generation through four-wave mixing in a single-mode optical fiber, Quantum Electronics, IEEE Journal of, 1987, 23(7): 1205-1210

[56] Cristiani, I., Tediosi, R., Tartara, L. & Degiorgio, V., Dispersive wave generation by solitons in microstructured optical fibers, Opt. Express, 2004, 12(1): 124-135

[57] Cheng, J.-X., Jia, Y. K., Zheng, G. & Xie, X. S., Laser-Scanning Coherent Anti-Stokes Raman Scattering Microscopy and Applications to Cell Biology, Biophysical Journal, 2002, 83(1): 502-509

[58] Cheng, J.-X.Xie, X. S., Coherent Anti-Stokes Raman Scattering Microscopy: Instrumentation, Theory, and Applications, The Journal of Physical Chemistry B, 2003, 108(3): 827-840

[59] Horie, K., Ushiki, H. & Winnik, F. M., The Interaction of Light with Materials II, Molecular Photonics: Fundamentals and Practical Aspects, 177-199

[60] Vallée, F.Bogani, F., Coherent time-resolved investigation of LO-phonon dynamics in GaAs, Physical Review B, 1991, 43(14): 12049

[61] Waltner, P., Materny, A. & Kiefer, W., Phonon relaxation in CdSSe semiconductor quantum dots studied by femtosecond time-resolved coherent anti-Stokes Raman scattering, Journal of Applied Physics, 2000, 88(9): 5268-5271

[62] Penzkofer, A., Laubereau, A. & Kaiser, W., High intensity Raman interactions, Progress in Quantum Electronics, 1979, 6(2): 55-140

[63] Loudon, R. *The quantum theory of light*. (Oxford university press, 2000).

[64] El-Diasty, F., Coherent anti-Stokes Raman scattering: Spectroscopy and microscopy, Vibrational Spectroscopy, 2011, 55(1): 1-37

[65] Portnov, A., Rosenwaks, S. & Bar, I., Detection of particles of explosives via backward coherent anti-Stokes Raman spectroscopy, Applied Physics Letters, 2008, 93(4): -

[66] Begley, R. F., Harvey, A. B. & Byer, R. L., Coherent anti-Stokes Raman spectroscopy, Applied Physics Letters, 1974, 25(7): 387-390

[67] Cui, M., Bachler, B. R. & Ogilvie, J. P., Comparing coherent and spontaneous Raman scattering under biological imaging conditions, Opt. Lett., 2009, 34(6): 773-775

[68] Zumbusch, A., Holtom, G. R. & Xie, X. S., Three-Dimensional Vibrational Imaging by Coherent Anti-Stokes Raman Scattering, Physical Review Letters, 1999, 82(20): 4142-4145

[69] Bergner, G., Chatzipapadopoulos, S., Akimov, D., Dietzek, B. *et al.*, Quantitative CARS Microscopic Detection of Analytes and Their Isotopomers in a Two-Channel Microfluidic Chip, Small, 2009, 5(24): 2816-2818

[70] Bergner, G., Albert, C. R., Schiller, M., Bringmann, G. *et al.*, Quantitative detection of C-deuterated drugs by CARS microscopy and Raman microspectroscopy, Analyst, 2011, 136(18): 3686-3693

[71] Zimmerley, M., Lin, C.-Y., Oertel, D. C., Marsh, J. M. *et al.*, Quantitative detection of chemical compounds in human hair with coherent anti-Stokes Raman scattering microscopy, BIOMEDO, 2009, 14(4): 044019-044019-044017

[72] Beeker, W. P., Groß, P., Lee, C. J., Cleff, C. *et al.*, A route to sub-diffraction-limited ? CARS Microscopy, Opt. Express, 2009, 17(25): 22632-22638

[73] Nikolaenko, A., Krishnamachari, V. V. & Potma, E. O., Interferometric switching of coherent anti-Stokes Raman scattering signals in microscopy, Physical Review A, 2009, 79(1): 013823

[74] Beeker, W. P., Lee, C. J., Boller, K.-J., Groß, P. *et al.*, Spatially dependent Rabi oscillations: An approach to sub-diffraction-limited coherent anti-Stokes Raman-scattering microscopy, Physical Review A, 2010, 81(1): 012507

[75] Tada, K.Karasawa, N., Broadband coherent anti-Stokes Raman scattering spectroscopy using soliton pulse trains from a photonic crystal fiber, Optics Communications, 2009, 282(19): 3948-3952

[76] Sidorov-Biryukov, D. A., Serebryannikov, E. E. & Zheltikov, A. M., Time-resolved coherent anti-Stokes Raman scattering with a femtosecond soliton output of a photonic-crystal fiber, Opt. Lett., 2006, 31(15): 2323-2325

[77] Paulsen, H. N., Hilligse, K. M., Thøgersen, J., Keiding, S. R. *et al.*, Coherent anti-Stokes Raman scattering microscopy with a photonic crystal fiber based light source, Opt. Lett., 2003, 28(13): 1123-1125

[78] Beaud, P., Hodel, W., Zysset, B. & Weber, H., Ultrashort pulse propagation, pulse breakup, and fundamental soliton formation in a single-mode optical fiber, Quantum Electronics, IEEE Journal of, 1987, 23(11): 1938-1946

[79] Aitchison, J., Weiner, A., Silberberg, Y., Oliver, M. *et al.*, Observation of spatial optical solitons in a nonlinear glass waveguide, Opt. Lett., 1990, 15(9): 471-473

[80] Gordon, J. P., Theory of the soliton self-frequency shift, Opt. Lett., 1986, 11(10): 662-664

[81] Mitschke, F. M.Mollenauer, L. F., Discovery of the soliton self-frequency shift, Opt. Lett., 1986, 11(10): 659-661

[82] Hooper, L. E., Mosley, P. J., Muir, A. C., Wadsworth, W. J. *et al.* in *Conference on Lasers and Electro-Optics.* CTuX4 (Optical Society of America).

[83] Hooper, L. E., Mosley, P. J., Muir, A. C., Wadsworth, W. J. *et al.*, Coherent supercontinuum generation in photonic crystal fiber with all-normal group velocity dispersion, Opt. Express, 2011, 19(6): 4902-4907

[84] Hartung, A., Heidt, A. M. & Bartelt, H., Design of all-normal dispersion microstructured optical fibers for pulse-preserving supercontinuum generation, Opt. Express, 2011, 19(8): 7742-7749

[85] Tomlinson, W., Stolen, R. H. & Johnson, A. M., Optical wave breaking of pulses in nonlinear optical fibers, Opt. Lett., 1985, 10(9): 457-459

[86] Dorrer, C., Salin, F., Verluise, F. & Huignard, J., Programmable phase control of femtosecond pulses by use of a nonpixelated spatial light modulator, Opt. Lett., 1998, 23(9): 709-711

[87] Karasawa, N., Li, L., Suguro, A., Shigekawa, H. *et al.*, Optical pulse compression to 5.0 fs by use of only a spatial light modulator for phase compensation, JOSA B, 2001, 18(11): 1742-1746

[88] Knoll, B.Keilmann, F., Near-field probing of vibrational absorption for chemical microscopy, Nature, 1999, 399(6732): 134-137

[89] Chabal, Y. J., Surface infrared spectroscopy, Surface Science Reports, 1988, 8(5): 211-357

[90] Conley, R. T. *Infrared spectroscopy*. (Allyn and Bacon, 1972).

[91] Chappell, J., Bloch, A., Bryden, W., Maxfield, M. *et al.*, Degree of charge transfer in organic conductors by infrared absorption spectroscopy, Journal of the American Chemical Society, 1981, 103(9): 2442-2443

[92] Zhu, X., Suhr, H. & Shen, Y., Surface vibrational spectroscopy by infrared-visible sum frequency generation, Physical Review B, 1987, 35(6): 3047

[93] Kahn, A. H., Theory of the infrared absorption of carriers in germanium and silicon, Physical Review, 1955, 97(6): 1647

[94] Iwata, K.Hamaguchi, H.-O., Construction of a Versatile Microsecond Time-Resolved Infrared Spectrometer, Appl. Spectrosc., 1990, 44(9): 1431-1437

[95] VAN DE WEERT, M., Fourier Transform Infrared, Methods for Structural Analysis of Protein Pharmaceuticals, 2005, 3(131

[96] Rochow, T. G.Tucker, P. A. in *Introduction to Microscopy by Means of Light, Electrons, X Rays, or Acoustics* 257-264 (Springer, 1994).

[97] Arrondo, J. L. R., Muga, A., Castresana, J. & Goñi, F. M., Quantitative studies of the structure of proteins in solution by Fourier-transform infrared spectroscopy, Progress in biophysics and molecular biology, 1993, 59(1): 23-56

[98] Helm, D., Labischinski, H., Schallehn, G. & Naumann, D., Classification and identification of bacteria by Fourier-transform infrared spectroscopy, Journal of general microbiology, 1991, 137(1): 69-79

[99] Pasquini, C., Near infrared spectroscopy: fundamentals, practical aspects and analytical applications, Journal of the Brazilian Chemical Society, 2003, 14(2): 198-219

[100] Gagel, J. J.Biemann, K., Continuous recording of reflection-absorbance Fourier transform infrared spectra of the effluent of a microbore liquid chromatograph, Analytical chemistry, 1986, 58(11): 2184-2189

[101] Buican, T. N.Carrieri, A. H. Ultra-high speed solid-state FTIR spectroscopy and applications for chemical defense. (DTIC Document, 2004).

[102] Schmitt, J.Flemming, H.-C., FTIR-spectroscopy in microbial and material analysis, International Biodeterioration & Biodegradation, 1998, 41(1): 1-11

[103] Pandey, K., A study of chemical structure of soft and hardwood and wood polymers by FTIR spectroscopy, Journal of Applied Polymer Science, 1999, 71(12): 1969-1975

[104] Meilunas, R. J., Bentsen, J. G. & Steinberg, A., Analysis of aged paint binders by FTIR spectroscopy, Studies in conservation, 1990, 35(1): 33-51

[105] Carr, G., Resolution limits for infrared microspectroscopy explored with synchrotron radiation, Review of Scientific Instruments, 2001, 72(3): 1613-1619

[106] Knoll, B.Keilmann, F., Enhanced dielectric contrast in scattering-type scanning near-field optical microscopy, Optics communications, 2000, 182(4): 321-328

[107] Hillenbrand, R., Knoll, B. & Keilmann, F., Pure optical contrast in scattering-type scanning near-field microscopy, Journal of microscopy, 2001, 202(1): 77-83

[108] Dazzi, A., Prazeres, R., Glotin, F. & Ortega, J. M., Local infrared microspectroscopy with subwavelength spatial resolution with an atomic force microscope tip used as a photothermal sensor, Opt. Lett., 2005, 30(18): 2388-2390

[109] Dazzi, A., Prazeres, R., Glotin, F. & Ortega, J., Subwavelength infrared spectromicroscopy using an AFM as a local absorption sensor, Infrared physics & technology, 2006, 49(1): 113-121

[110] Dazzi, A., Prazeres, R., Glotin, F. & Ortega, J., Analysis of nano-chemical mapping performed by an AFM-based ("AFMIR") acousto-optic technique, Ultramicroscopy, 2007, 107(12): 1194-1200

[111] Holman, H. Y. N., Martin, M. C., Blakely, E. A., Bjornstad, K. *et al.*, IR spectroscopic characteristics of cell cycle and cell death probed by synchrotron radiation based Fourier transform IR spectromicroscopy, Biopolymers, 2000, 57(6): 329-335

[112] Yuan, W., Coherent broadband mid-infrared supercontinuum generation in As2Se3 photonic crystal fiber, arXiv preprint arXiv:1309.5830, 2013,

[113] Shaw, L., Thielen, P., Kung, F., Nguyen, V. *et al.* in *Conf. Adv. Solid State Lasers (ASSL), Seattle, WA.*

[114] Price, J. H. V., Feng, X., Heidt, A. M., Brambilla, G. *et al.*, Supercontinuum generation in non-silica fibers, Optical Fiber Technology, 2012, 18(5): 327-344

[115] Kubat, I., Agger, C. S., Moselund, P. M. & Bang, O., Mid-infrared supercontinuum generation to 4.5 μm in uniform and tapered ZBLAN step-index fibers by direct pumping at 1064 or 1550 nm, JOSA B, 2013, 30(10): 2743-2757

[116] Granzow, N., Stark, S. P., Schmidt, M. A., Tverjanovich, A. *et al.*, Supercontinuum generation in chalcogenide-silica step-index fibers, Opt. Express, 2011, 19(21): 21003-21010

[117] Baili, A., Cherif, R., Saini, T. S., Kumar, A. *et al.* 92000S-92000S-92006.

[118] Ung, B.Skorobogatiy, M., Chalcogenide microporous fibers for linear and nonlinear applications in the mid-infrared, Opt. Express, 2010, 18(8): 8647-8659

[119] Hu, J., Menyuk, C. R., Shaw, L. B., Sanghera, J. S. *et al.* in *Lasers and Electro-Optics, 2008 and 2008 Conference on Quantum* Electronics and Laser Science. CLEO/QELS 2008. Conference on. 1-2 (IET).

Chapter 3

Luminescence-Spectrum Modification of White Light-Emitting Diodes by Using 3D Colloidal Photonic Crystals

Chun-Feng Lai

Additional information is available at the end of the chapter

http://dx.doi.org/10.5772/59934

1. Introduction

1.1. General background of white light-emitting diodes

In recent times, light-emitting diodes (LEDs) have been used all over the world, such as general lighting, automotive lighting, backlighting of displays, and street signals. In addition, white LEDs (WLEDs) that have grown rapidly have been a popular source of general illumination. WLEDs have attracted considerable attention for the lighting industry because of their numerous advantages, such as high brightness, high reliability, and low energy consumption [1]. The color of WLEDs is evaluated by two parameters: the correlated color temperature (CCT) and the color-rendering index (CRI). WLEDs used for general lighting exhibit CCT of approximately 2700 K (as incandescent bulbs), 3000 K (as warm-fluorescent tubes), and 5000 K (as cold-fluorescent tubes), respectively. Cold-WLEDs (*c*-WLEDs) are used for general lighting, and warm-WLEDs (*w*-WLEDs) create a feeling of wellbeing. The *w*-WLEDs are a particularly valuable light source because they provide comfortable residential lighting, is approximately 2700 K to 3000 K.

Several approaches to fabricate WLEDs have been developed [2-3]. Commercially WLEDs generally use a GaN blue-emitting LED for stimulating yellow-emitting phosphor (YAG:Ce^{3+}) to yield phosphor-converted WLEDs, which are the most efficient lights that provide high luminous efficiency and a low cost [4]. Although phosphor-converted WLEDs have numerous advantages, for example low cost, and high phosphor conversion efficiency. However, several PC-WLEDs problems remain, including a high CCT and a low CRI [5-6]. All WLEDs have required achieving a CRI that exceeds 80, a standard that is used in general lighting applications. Therefore, two or more color phosphors are required when the CRI exceeds 90. That have been achieved by WLEDs with a larger weight percent (wt%) of red phosphor, or by multi-color LED combinations [7]. Currently, all nitride- and borate-based red phosphors are typically expensive than yellow phosphors, a phenomenon that contributes to the high cost of fabrication *w*-WLEDs. Although WLEDs are brighter that compared to the traditional incandescent bulbs. Until now, many people still do not accept that the expensive WLED bulbs. Expanding the WLED illumination market is difficult due to the high cost of WLEDs. Reducing the cost of WLEDs can hasten the transition from conventional bulbs to LEDs and lead to new applications in optical devices.

1.2. General background of photonic crystals

Photonic crystals (PhCs) were first proposed by E. Yablonovitch [8] and S. John [9] in 1987. They proposed a period arrangement of refractive index can possess the photonic bandgap (PBG) in the wavelength range. PBG is, photons can only travel across the material if they are localized into distinct energy states and obey strict rules relating to direction of travel, polarization state and wavelength. A wavelength range can also exist for which there are no allowed states for propagation. PhCs are regular arrays of materials with different refractive indexes that are classified into three main categories, according to the dimensionality of the stack, such as one- (1D), two- (2D), and three-dimensional (3D), as shown in Figure 1. PhCs nanostructures provide new ways to control photons. Recently, optical properties of the PhCs have been much study. 1D PhCs have been applied on many technology, such as Bragg reflectors of the optical feedback mechanism in distributed feedback lasers [10] and vertical cavity surface emitting lasers [11]. In addition, 2D and 3D PhCs has been subject of much intensive research in areas related to LEDs [12-14], sensing [15], telecommunications [16], slow light [17], and quantum optics [18].

In this chapter, we focus on the PBG material of 3D colloidal PhCs (CPhCs), which is important because it exhibits a forbidden optical energy band [19]. 3D CPhCs have been studied extensively because of the unique optical properties and applications [20]. The refractive index of colloidal particles gives rise to PBG properties. Light energy in the PBG cannot propagate through materials, and the light is consequently reflected. The 3D CPhC structures, that based on the face-centred-cubic (fcc) opals and inverse opals, are interesting due to they can be produced using colloidal solutions.

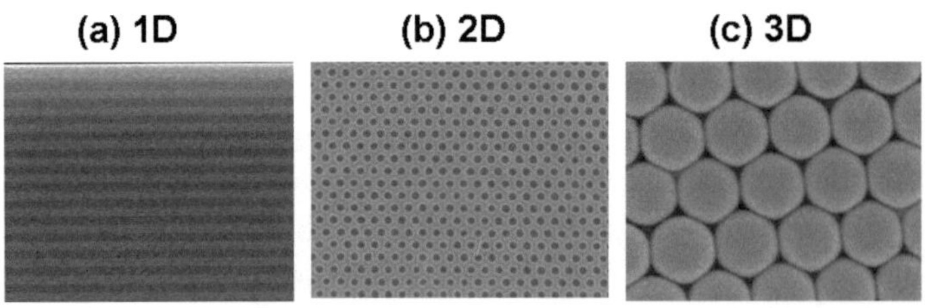

Figure 1. Schematic illustration of PhCs with periodicity in (a) 1D, (b) 2D, and (c) 3D.

1.3. Research niche

WLEDs have multiple advantages, such as their small size, conservation of energy, and long lifetime. WLEDs will comprehensive to replace conventional lighting sources within years.

Commercial WLEDs are made by a GaN-based blue LED combination with a yellow-emitting YAG:Ce phosphor; the combination of blue and broadband yellow approximates white light. For residential lighting, many people prefer the *w*-WLEDs; to achieve this, the proportion of light emitted by the red-emitting phosphor to the mix. Thus, *w*-WLEDs have lower efficiency than *c*-WLEDs. We have developed a novel approach for improving color quality: layering the 3D CPhC with different diameter (*D*) over the phosphor layer. The PBGs in the 3D CPhC improve both the color quality (quantified as the CRI) and the closeness to a blackbody emission (quantified as the CCT).

In this chapter, we first introduce the theoretical analysis and synthesis method of 3D CPhC structures in section 2. Then, in section 3, we explain the fabrication method of WLEDs and 3D CPhCs, where 3D CPhCs includes polystyrene (PS) nanospheres and silica (SiO_2) nanospheres. Next, in section 4, we demonstrate the theoretically and experimentally for luminescence-spectrum modification of WLEDs by using 3D CPhCs. Finally, conclusions provide in section 5.

2. Fundamental and modelling of 3D colloidal photonic crystals

2.1. Photon band structure properties of 3D colloidal photonic crystal structures

Three-dimensional (3D) PhC, which can have the novel properties, is periodic along three different axes. For 3D PhCs, complete PBG are more rare. The PBG must smother the entire 3D Brillouin zone, not just any one plane or line. For example, Figure 2 shows the photonic band structure (PBS) for an fcc lattice of close-packed PS nanospheres (n_{PS}=1.59) in air (n_{air}=1.0). There is no complete PBG. Nevertheless, a number of 3D PhCs have been discovered that do yield sizable complete PBGs [21-24]. In most of the theoretical studies of these structures undertaken to date, the results are as follows. For a given lattice constant of 3D PhCs, there is no PBG until the refractive index of the material is increased to some value. The PBG opens up and its width usually increases monotonically with the refractive index of the material.

In Figure 2, we showed that an fcc lattice of nanospheres does not have a complete PBG [24]. Sanders found that precious opal mineraloids are formed of close-packed arrangements of submicron-diameter silica spheres in a silica–water matrix [25], with a relatively low dielectric contrast. Just as for the case of an fcc lattice of close-packed dielectric spheres (see Figure 2(b)), small gaps appear only at particular points in the band diagram. The wavevectors *k* of these partial gaps corresponds to particular directions at which a particular wavelength, and therefore a particular colour, is reflected. The narrowness and directionality of these gaps are the source of the bright, iridescent colours that make opal gems so attractive. This sort of structural colour (as opposed to colours that results from absorption by chemical pigments) is responsible for many of the iridescent colours found in nature, from butterfly wings, to peacock feathers, to certain beetles, and jellyfish. In this chapter, we focus on a synthetic opal that is similar to natural opals and can be readily fabricated, because microscopic spheres can be induced to self-assemble into an fcc lattice as one evaporates a solution in which they are suspended (a colloid).

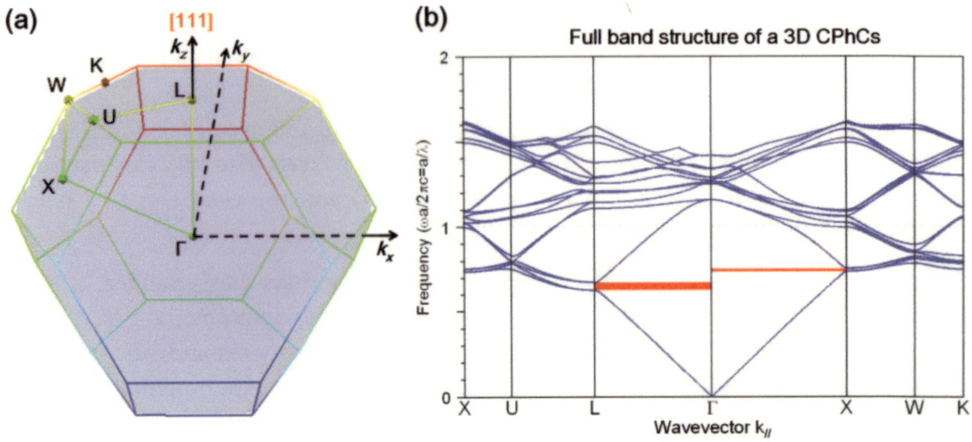

Figure 2. (a) Schematic illustration of the fcc Brillouin zone. (b) The PBS for the lowest-frequency electromagnetic modes of an fcc lattice of close-packed polystyrene (PS) nanospheres (n_{PS}=1.59) in air.

2.2. Bragg's law analysis methods for 3D colloidal photonic crystals

The 3D CPhCs are artificial opal structures of dielectric materials that exhibit unique photonic dispersion properties and that control light emission behaviours. The optimal design of 3D CPhC structures is strongly dependent on three parameters, for example lattice constant (*a*), filling factor (*f*), and refractive index (*n*). To discuss the effect of the lattice constant *a*, the Bragg's reflection theory is used [26]. 3D CPhCs are known to possess PBG properties; therefore, during the performance of normal reflectance spectrum measurements, the incident light direction was assumed to be perpendicular to the (111) plane ($\theta_{111}=0$), as shown in Figure 3. The reflection wavelength (λ_R) of the Bragg law was calculated expressed as

$$\lambda_R = 2d_{111}\sqrt{n_{eff}{}^2 - \sin^2\theta_{111}} \tag{1}$$

where λ_R is the reflection wavelength, d_{111} is the inter-planar spacing between the (111) planes, and n_{eff} is the effective refractive index. For the 3D CPhC structures, n_{eff} can be approximated as follows:

$$n_{eff}{}^2 = n_{PS}{}^2 f_{PS} + n_{air}{}^2\left[1 - f_{PS}\right] \tag{2}$$

where n_{PS} = 1.59 and n_{air} = 1.0 are the refractive indices of the PS nanospheres and air, respectively. The monodispersed colloidal nanospheres typically adopted an fcc packing to form the nanosphere volume fraction (f_{PS}), which was 74 vol% in the structure [27]. This adhered to the measured reflection results, as shown in Figure 7 and Figure 8, and indicates that light waves cannot propagate within this region.

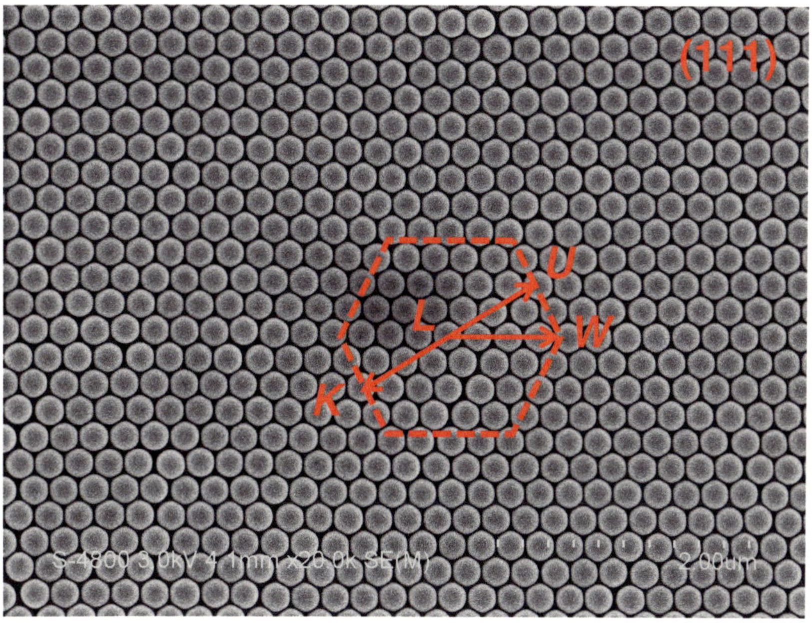

Figure 3. The field-emission scanning electron microscope images of the 3D CPhCs show that the PS nanospheres stack with a well-organized (111) plane. Inset: the first Brillouin zone with the symmetry points.

3. Fabrication method of white light-emitting diodes and 3D colloidal photonic crystals

3.1. White light-emitting diodes prepared

The *c*-WLEDs (CCT of 5283 K) and *w*-WLEDs (CCT of 3130 K) were fabricated with GaN blue-emitting LED chips and varying amounts of yellow- and red-emitting phosphor mixed with a silicone encapsulant. Figure 4 shows the 3D CPhCs deposited onto the plastic leaded chip carrier (PLCC) WLED package. The *c*-WLEDs and *w*-WLEDs were fabricated as follows [12-13]: First, we bonded a 22 × 35 mil^2 chip of GaN pattern-sapphire substrate (PSS) LEDs (emission wavelength, λ = 455 nm) was bonded to a commercial PLCC package using silver paste and gold wire. The GaN blue-emitting LED structures were grown on the PSS by using a metal organic chemical vapour deposition (MOCVD). The GaN blue-emitting LED structure consisted of a 30-nm-thick AlN buffer layer, a 2-μm-thick undoped GaN layer, a 2-μm-thick n-type GaN layer, a 100 nm InGaN/GaN multiple quantum well active region, and a 0.1-μm-thick p-type GaN layer. In addition, a 200-nm-thick indium-tin-oxide layer was deposited on the p-GaN layer to form a transparent contact layer (TCL). Finally, the Cr/Pt/Au (5/5/1000 nm) metal layers were deposited as n-type and p-type contact pads. The layout of the GaN blue-emitting PSS LED chip is shown in Figure 4. Second, we used $Y_3Al_5O_{12}$:Ce^{3+} (yellow-emitting) and $Ba_2Si_5N_8$:Eu^{2+} (red-emitting) phosphors to obtain the *c*-WLEDs and *w*-WLEDs. The yellow- and red-emitting phosphors (of various wt%) were uniformly mixed with the silicone, and then filled into the lead-frame by using phosphor dispensing techniques. The *c*-WLEDs had a yellow-emitting and red-emitting phosphor concentration of 16.8 wt% and 1.2 wt%, respectively, and the *w*-WLEDs had a yellow-emitting and red-emitting phosphor concentration of 21.0 wt% and 4.0 wt%, respectively. In this study, increasing the concentration of the yellow-emitting phosphor in the *c*-WLEDs by 125% and the red-emitting phosphor concentration by 300% caused the high CCT of the *c*-WLEDs to drop to the low CCT of the *w*-WLEDs.

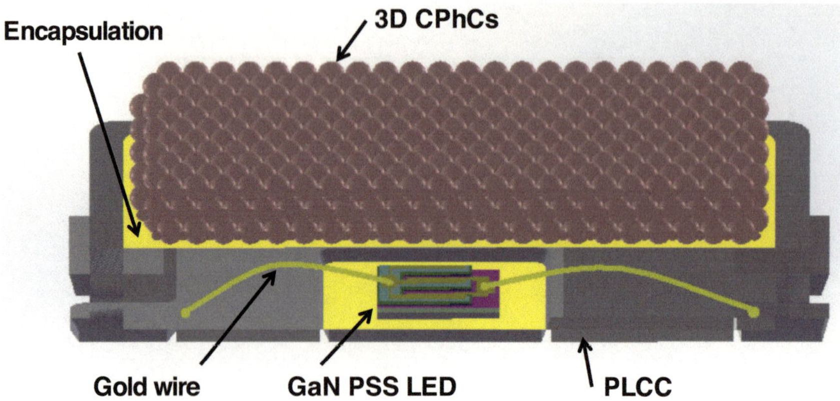

Figure 4. Schematic diagram of the WLED PLCC package with the 3D CPhCs.

3.2. Polystyrene nanosphere of 3D colloidal PhCs prepared

PS nanospheres were manufactured as follows. The monodisperse polymer nanospheres were prepared using styrene (Acros Organics) as the monomer, sodium dodecyl sulphate (SDS; Acros Organics) as the emulsifier, and potassium persulphate (KPS; Acros Organics) as the initiator in emulsion polymerization [28]. In this synthesis process, 15 mL of styrene, 100 mg of SDS, and 200 mL of deionized (DI) water were added to a 1000 mL three-necked flask, which was placed in a water bath at 70 °C in an atmosphere of nitrogen gas. Subsequently, 250 mg of KPS dissolved in 50 mL of DI water was added to the mixture while stirring at 300 rpm. After stirring for 24 h, monodispersed PS nanospheres with an average diameter of 270 nm were obtained. By varying the amount of styrene monomer, PS colloidal spheres with different sizes were synthesized with the same method. The 10-μL 3D PS CPhCs were deposited on the entire emission region of the *w*-WLEDs through natural sedimentation in a 60 °C oven. A field-emission scanning electron microscope (FESEM) was used to study the crystalline structure of the 3D PS CPhCs produced on the WLEDs. The 3D PS CPhCs with fcc structures were grown through natural sedimentation, with their surfaces parallel to the (111) crystallographic plane. We prepared four sizes of the PS nanosphere, with *D* of 190 nm, 220 nm, 230 nm, and 250 nm, as shown in Figure 5. These images indicate that the 3D PS CPhC nanostructure consists of an fcc array. In this study, 3D PS CPhCs with high-reflectivity were obtained at a growth temperature of 60 °C. The average 10-μm of 3D PS CPhC thin-film was prepared on the WLED surface, indicating that the practical 3D PS CPhC process did not suffer adverse effects according to the experimental results. In addition, the normal reflection spectrum for this 3D

PS CPhCs showed the different reflection peak positions, respectively, which adhered to the Bragg law (Figure 6).

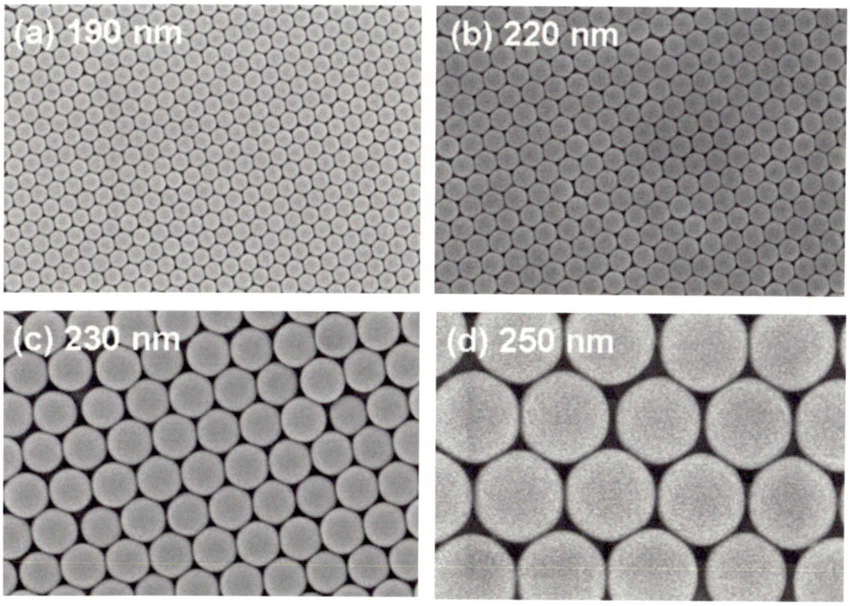

Figure 5. FESEM images of the 3D CPhCs fabricated by using PS nanospheres with *D* of (a) 190 nm, (b) 220 nm, (c) 230 nm, and (d) 250 nm.

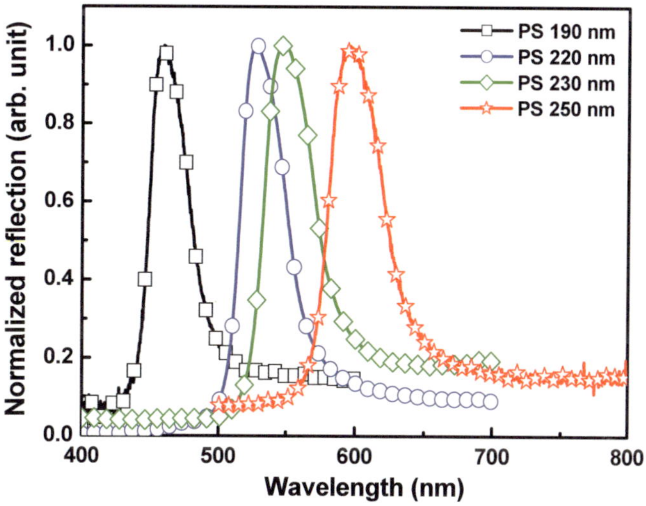

Figure 6. The reflection spectra were measured by the different PS nanosphere size of 3D CPhCs, respectively.

3.3. Silica nanosphere of 3D colloidal PhCs prepared

Silica nanospheres were synthesized using a modified Stober method [29] as follows. Monodispersed silica nanospheres were synthesized using ammonium hydroxide (NH_4OH; SHOWA), the tetraethoxysilane (TEOS; Aldrich), and the DI water condensation was controlled in anhydrous ethanol solution [20-31]. In this synthesis process, 200 mL of anhydrous ethanol, 10 mL of NH_4OH, and 50 mL of DI water were added to a 500 mL three-necked flask, which was placed in a water bath at 35 °C. Subsequently, 15 mL of TEOS was dropped to the mixture while stirring at 300 rpm. After stirring for 24 h, monodispersed silica nanospheres with an average diameter of 250 nm were obtained. By varying the amount of TEOS, silica colloidal nanospheres with different sizes were synthesized with the same method [32-33]. We prepared the latex (100.0 mg/mL) with silica nanosphere of four diameters (D = 150 nm, 180 nm, 250 nm, and 290 nm). The 3D silica CPhCs (10.0-μL) were deposited onto the surface of the WLEDs by using a vertical deposition method. Figures 7 show FESEM images of the (111) face of the 3D silica CPhCs. In addition, the normal reflection spectrum for this 3D silica CPhCs showed the different reflection peak positions, respectively, which conformed to the Bragg law (Figure 8).

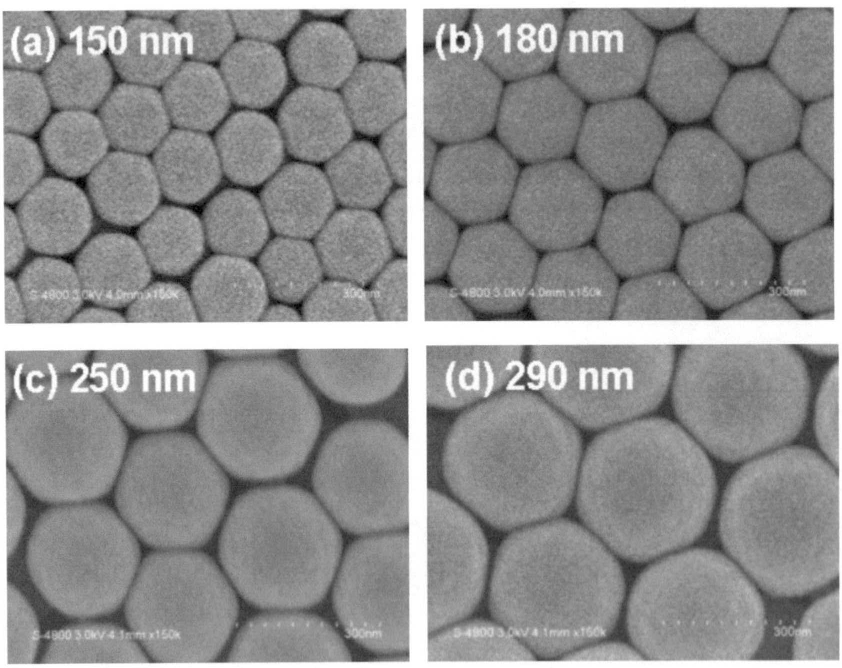

Figure 7. The FESEM images of the 3D CPhCs fabricated by silica nanospheres with D of (a) 150 nm, (b) 180 nm, (c) 250 nm, and (d) 290 nm.

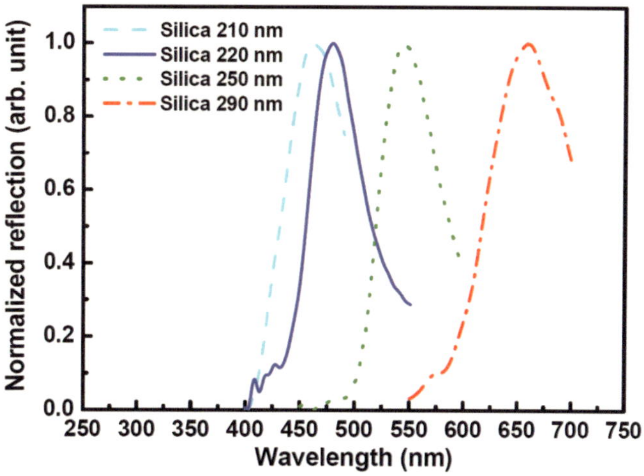

Figure 8. The reflection spectra were measured by the different silica nanosphere size of 3D CPhCs, respectively.

4. Luminescence-spectrum modification of WLEDs by using 3D colloidal photonic crystals

4.1. Luminescence spectra modification of WLEDs by using 3D PS CPhCs

The luminescence spectra of the WLEDs with and without 3D PS CPhCs were measured by the integration sphere at a current of 120 mA, as shown in Figure 9. Due to the photonic stop-bands, and phosphor reabsorption and reemission affect the luminescence spectrum of WLEDs with CPhCs [12]. The light emission of the WLEDs through the 3D PS CPhCs was propagated according to the PBSs of the 3D PS CPhCs. This study measured the angular-resolved transmission spectra to study the light emission distribution of the WLEDs with 3D PS CPhCs, which will be discussed in the next section.

In addition, we also measured the luminous flux, luminous efficiency, CRI, CCT, and the International Commission on Illumination (CIE) color chromaticity coordinates (x, y) versus current characteristics by using a integration sphere. The luminous flux values of the *c*-WLEDs, the *c*-WLEDs with 3D PS CPhC of $D = 230$ nm, the *c*-WLEDs with 3D PS CPhC of $D = 250$ nm, and the *w*-WLEDs were 38, 34, 31, and 35 lm, respectively, and the luminous efficiencies were 100, 90, 81, and 93 lm/W, respectively, at a input power of 0.38 W. The CCT of *c*-WLEDs with 3D PS CPhCs can decrease from approximately 5283 to approximately 3130 K without increasing the wt% of yellow- and red-emitting phosphor. In addition, the *c*-WLEDs with 3D PS CPhC of $D = 230$ nm demonstrated a luminous flux, CRI, and CCT equal to that of the reference *w*-WLEDs. The *c*-WLEDs containing 3D PS CPhC of $D = 250$ nm shows the highest CRI value of over 90. The CIE (x, y) coordinates of the four types of WLEDs fall on the Planckian locus. Therefore, this novel technology could reduce the fabrication cost of *w*-WLEDs.

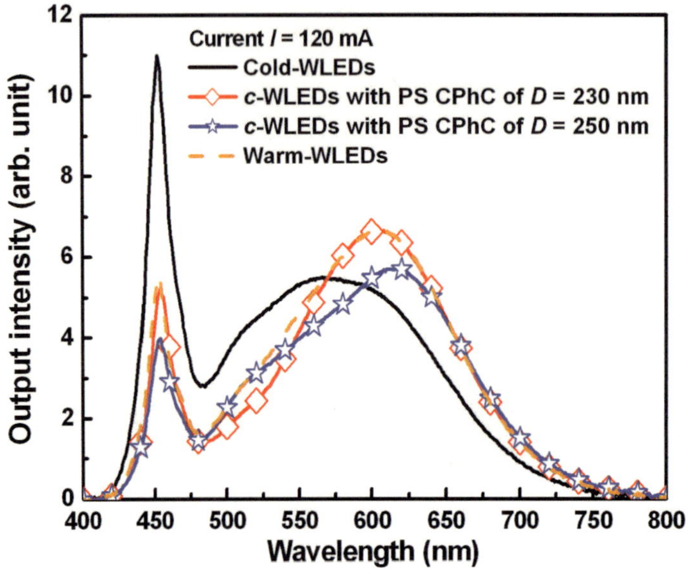

Figure 9. Luminescence spectra of the WLEDs.

4.2. Photon band structure theoretical discussion of WLEDs containing 3D PS CPhCs

Several research groups have investigated the optical properties of 3D CPhCs [34-37]. They have indicated a photonic stop-band at the *L* symmetry point of the fcc Brillouin zone along the [111] direction. The photonic stop-bands of the 3D PS CPhCs can be controlled by the nanosphere sizes. The luminescence spectra of the WLEDs were modified by 3D PS CPhCs, as shown in Figure 9. Figure 9 exhibited the CCT, CRI, and CIE (x, y) coordinates of the WLEDs that could be controlled by the emission intensity ratio of blue-, green-, yellow-, and red-emitting light. Generally, In this study, the luminous flux, CRI, CCT, and CIE (x, y) coordinates of the 3D PS CPhCs deposited onto the WLEDs were affected by PBSs depending on the lattice constant of 3D PS CPhCs.

The light emission distribution of WLEDs containing 3D PS CPhCs was measured using the angular-resolved transmission spectrum technique. The apparatus for the angular-resolved transmission spectra measurement was the same as in *Lai* et al. [38]. The *c*-WLEDs containing 3D PS CPhC of $D = 230$ nm and 250 nm measured on a current of 120 mA. The angular-resolved transmission spectra as a function of detection angle θ_{out} were obtained with a fibre probe coupled to an spectrometer (Horiba Jobin Yvon, CP140), which was rotated from $\theta_{out} = 0°$ (normal direction) until ±90° in increments of 0.5°. The measuring plane of the *c*-WLEDs containing 3D PS CPhCs was fixed along the *Γ-L-U* path of the fcc Brillouin zone, as indicated in Figure 3. The transmission spectra were displayed on a wavelength versus detection angle plot, and the color representing the intensity according to a log scale bar. Figures 10(a) and 10(b) show the transmission spectra collected along the *Γ-L-U* directions for the *c*-WLEDs containing 3D PS CPhC of $D = 230$ nm and 250 nm, respectively. These measurement results showed that the photonic stop-bands for the guided modes shifted toward shorter wavelengths with increasing detection angles θ_{out}. Figures 10(a) and 10(b) demonstrated the first photonic pseudogap of the PBSs along the (111) surface. Additionally, due to the photonic stop-bands caused the light diffracted back into the phosphor layer of *c*-WLEDs, which could be reabsorbed and reemitted by the phosphor layer to increase the red light emission, as shown in Figure 9.

The *c*-WLEDs containing 3D PS CPhCs also exhibited photonic stop-bands in the zenithal (θ) and azimuthal (φ) direction [12]. Figures 10(a) and 10(b) can be represented as the photonic dispersion curves with $k_{//} = a\ \sin(\theta_{out})/\lambda$ for comparison with the calculated PBSs [38]. These experimental results demonstrated a PBG that consisted of the lowest bands of the PBS in the *L-U* direction. The photonic stop-band corresponded to the *L-U* of the PBSs associated with Bragg diffraction by the (111) planes. We took the *Γ-L-U* triangle as representative, and assumed that the incident light with wave vector *k* was in that diffraction plane; with the tip of *k* lying on the *L-U* segment. The azimuthal of the far-field emission distribution has measured as a function of the azimuthal angle φ by using the angular-resolved transmission-measuring apparatus [38]. In azimuthal measurement results of *c*-WLEDs containing 3D PS CPhCs also showed the photonic stop-bands due to the PBS of the 3D PS CPhCs [12].

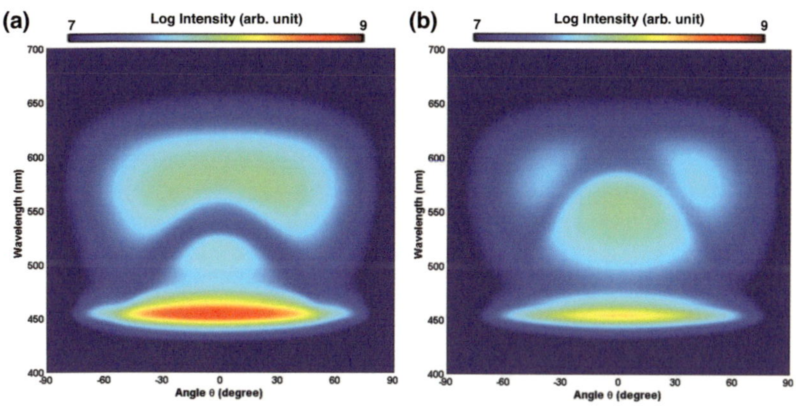

Figure 10. (a) and (b) Angular-resolved measurements of the *c*-WLEDs containing 3D PS CPhC of $D = 230$ nm and 250 nm.

4.3. Luminescence spectra modification of WLEDs by using 3D silica CPhCs

In this section, we prepared a silica nanosphere of diameter (D) of 400 nm. The normal reflection spectrum for this 3D silica CPhC showed a reflection peak at 923 nm that conformed to the Bragg law. 3D silica CPhC was deposited onto the WLEDs by a vertical deposition method. The 3D silica CPhC thin films modified the luminescence spectra of the WLEDs, which were measured using a 20-in integration sphere with a current setting of 120 mA, as shown in Figure 11; the 425–600-nm of light emission wavelengths of the WLEDs containing 3D silica CPhCs were suppressed by the 3D silica CPhC thin films. In addition, the red-light wavelength exceeding 600 nm of the WLEDs containing 3D silica CPhCs increased (Figure 11) because of the reabsorption and re-emission produced by the phosphor layer of the WLEDs [12-14].

In this study, we also measured the optical characteristics of WLEDs using the integration sphere that was equipped with a radiometer and photometer. The luminous flux of the WLEDs, *w*-WLEDs, and WLEDs containing 3D silica CPhCs were 53, 44, and 46 lm with an input power setting of 0.38 W, respectively. The luminous efficacy of these WLEDs was 144, 120, and 127 lm/W, respectively. The luminous flux of the WLEDs containing 3D silica CPhCs was 5.6% more enhanced than that of the *w*-WLEDs. Additionally, the CCT of WLEDs containing 3D silica CPhCs decreased from approximately 3974 K to 2960 K without an increase in the wt% of the green- and red-emitting phosphors. The CIE (x, y) coordinates of all of the light sources in this study were located near the Planckian locus of the CIE chromaticity diagram. All WLEDs must have a CRI that exceeds 80, a standard that is required for use in residential lighting applications.

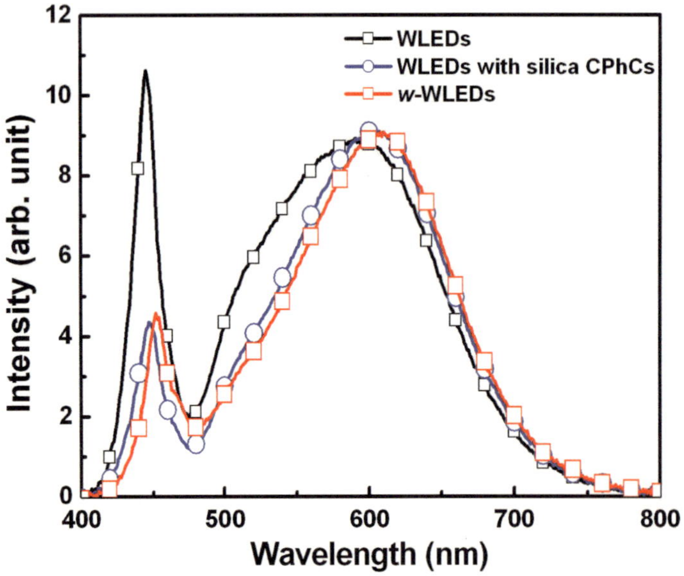

Figure 11. Luminescence spectra of the WLEDs.

4.4. Photon band structure theoretical discussion of WLEDs containing 3D silica CPhCs

The angular-resolved reflection spectra were measured to examine the light-reflection distribution of the 3D silica CPhCs; Figure 12 shows the PBS of the 3D silica CPhCs. The angular-resolved reflection measurement systems used in this study were identical to those described previously [32-33]. To confirm that the high-order band of PBS affected the luminescence spectrum of the WLED, we measured the transmittance spectrum of the 3D silica CPhC thin films by using a UV-Vis spectrophotometer (Figure 13). We used a transmittance measurement system equipped with a halogen lamp as the white-light source, an integration sphere, and a spectrometer with a charge-coupled device to measure the transmittance spectra through the 3D silica CPhC thin films. The 3D silica CPhC thin films exhibited a graduated transmission from the blue wavelength region to the red wavelength region. The transmittance of 3D silica CPhC thin films was similar to that of the graduated neutral density filter (GNDF). The 400–600-nm wavelength spanning from the blue region to the red region exhibited a low transmittance that was affected by the high-order band of PBS (Figure 13). In other words, when the WLED photons passed through the 3D silica CPhC thin films, one part was reflected and the other was transmitted. The reflected light of the WLEDs can be absorbed by the phosphor layer and increase the red-light emission [12-14]. In this section, we report that the WLED devices containing 3D silica CPhCs achieved a higher luminous flux than did the commercial *w*-WLEDs, because of the GNDF influence and increased light re-emission in the red-light region.

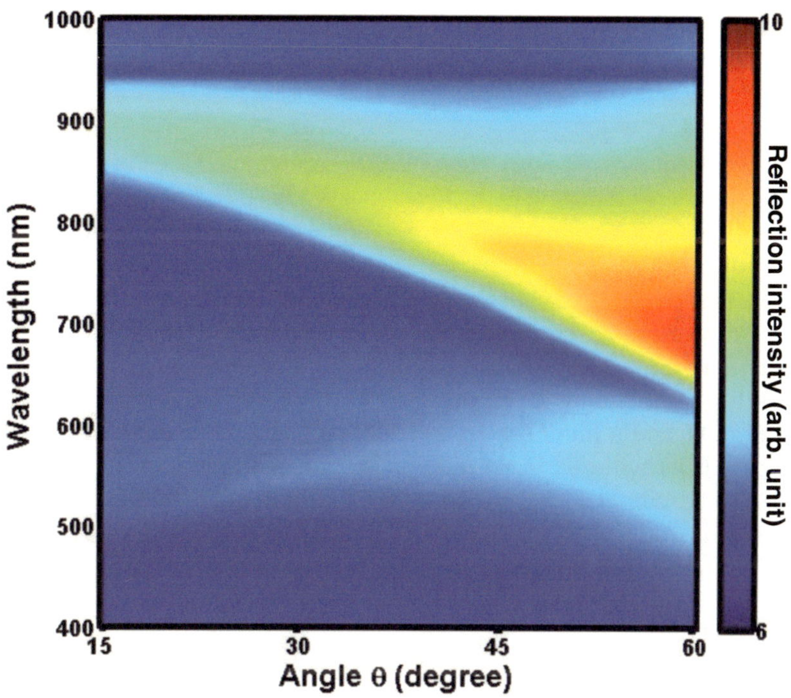

Figure 12. Angular-resolved reflection measurements of 3D silica CPhC thin films.

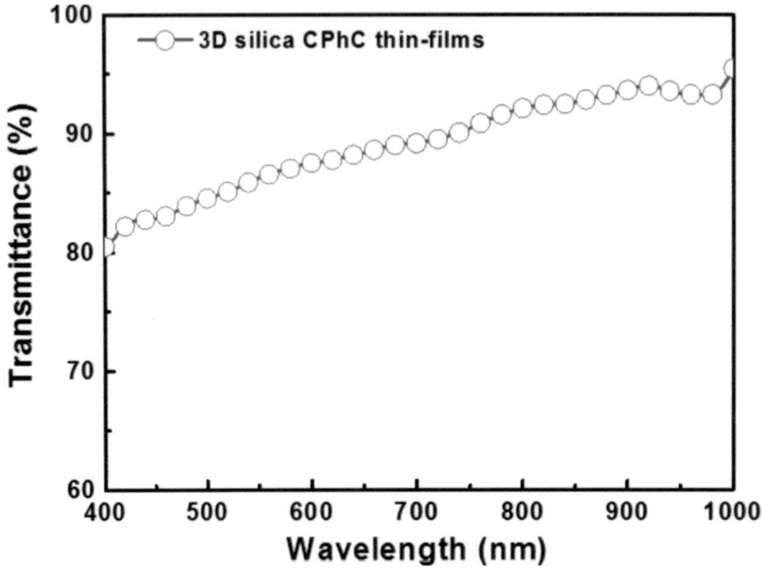

Figure 13. Measured transmittance of 3D silica CPhC thin films.

5. Conclusion

In conclusion, we designed and fabricated WLEDs containing 3D CPhCs that had luminescence modified that were according to the PBSs of 3D CPhCs. The *c*-WLEDs with 3D PS CPhC of $D = 230$ nm shows the optical properties almost equal to that of the reference *w*-WLEDs. The angular-resolved transmission measurement showed the PBGs that attributed to the PBSs of the 3D PS CPhCs. The *c*-WLEDs containing 3D PS CPhCs caused the shorter wavelength of light emission to reemission the long wavelength because of the PBSs. The experimental transmission spectra agreed favourably with the simulated PBSs of the 3D PS CPhCs. In addition, the WLEDs containing 3D silica CPhCs in 120 mA produced a CCT of 2960 K, and exhibited a CRI of 80 and a luminous flux of 46.7 lm (5.6% more enhanced than that of the commercial *w*-WLEDs) without an increase in the concentration of the phosphors. In the WLEDs containing 3D silica CPhCs, the 3D silica CPhC thin films modified the light emission spectrum because of the GNDF influence and multiple phosphor re-emissions. This novel technique applied to WLEDs to produce *w*-WLEDs reduces fabrication costs.

Acknowledgements

The authors gratefully acknowledge the financial support for this research by the National Science Council (NSC), and Ministry of Science and Technology (MOST) in Taiwan, under grant numbers NSC102-2221-E-035-046, NSC102-2622-E-035-030-CC2, MOST103-2221-E-035-029, and MOST103-2622-E-035-007-CC2.

Author details

Chun-Feng Lai

Address all correspondence to: chunflai@fcu.edu.tw

Department of Photonics, Feng Chia University, Taichung, Taiwan

References

[1] Schubert, E. F., and Kim, J. K. (2005). Solid-state light sources getting smart. *Science*, Vol. 308, pp. 1274-1278.

[2] Kwak, J., Bae, W. K., Lee, D., Park, I., Lim, J., Park, M., Cho, H., Woo, H., Yoon, D. Y., Char, K., Lee, S., and Lee, C. (2012) Bright and efficient full-color colloidal quantum dot light-emitting diodes using an inverted device structure. *Nano lett.*, Vol. 12, pp. 2362-2366.

[3] Nizamoglu, S., Ozel, T., Sari, E., and Demir, H. V. (2007) White light generation using CdSe/ZnS core–shell nanocrystals hybridized with InGaN/GaN light emitting diodes. *Nanotechnology*, Vol. 18, 065709.

[4] Steigerwald, D. A., Bhat, J. C., Collins, D., Fletcher, R. M., Holcomb, M. O., Ludowise, M. J., Martin, P. S., and Rudaz, S. L. (2002) Illumination with solid state lighting technology. *IEEE J. Sel. Top. Quant. Electron.*, Vol. 8, pp. 310-320.

[5] You, J. P., Tran, N. T., and Shi, F. G. (2010) Light extraction enhanced white light-emitting diodes with multi-layered phosphor configuration. *Opt. Express*, Vol. 18, pp. 5055-5060.

[6] Park, J. K., Choi, K. J., Yeon, J. H., Lee, S. J., and Kim, C. K. (2006) Embodiment of the warm white-light-emitting diodes by using a Ba^{2+} codoped Sr_3SiO_5:Eu phosphor. *Appl. Phys. Lett.*, Vol. 88, 043511.

[7] Oh, J. H., Yang, S. J., Sung, Y. G., and Do, Y. R. (2012) Excellent colour rendering indexes of multipackage white LEDs. *Opt. Express*, Vol. 20, pp. 20276-20285.

[8] Yablonovitch, E. (1987) Inhibited spontaneous emission in solid-state physics and electronics. *Phy. Rev. Lett.*, Vol. 58, pp. 2059.

[9] John, S. (1987) Strong localization of photons in certain disordered dielectricsuper lattices. *Phys. Rev. Lett.*, Vol. 58, pp. 2486-2489.

[10] Kogelnik, H., and Shank, C. V. (1971) Stimulated emission in a periodic structure. *Appl. Phys. Lett.*, Vol. 18, pp. 152-154.

[11] Soda, H., Iga, K., Kitahara, C., and Suematsu, Y. (1979) GaInAsP/InP surface emitting injection lasers. *Jap. J. Appl. Phys.*, Vol. 18, pp. 2329-2330.

[12] Lai, C. F., Chang, C. C., Wang, M. J., and Wu, M. K. (2013) CCT- and CRI-tuning of white light-emitting diodes using three-dimensional non-close-packed colloidal photonic crystals with photonic stop-bands. *Opt. Express*, Vol. 21, pp. A687-A694.

[13] Lai, C. F., Hsieh, C. L., and Wu, C. J. (2013) Light-spectrum modification of warm white-light-emitting diodes with 3D colloidal photonic crystals to approximate candlelight. *Opt. Letters*, Vol. 38, pp. 3612-3615.

[14] Lai, C. F., Lee, Y. C., and Kuo, C. T. (2014) Saving phosphor by 150% and producing high color-rendering index candlelight LEDs containing composite photonic crystals. *J. Light. Technol.*, Vol. 32, pp. 1930-1935.

[15] Loncar, M., Scherer, A., and Qiu, Y. (2003) Photonic crystal laser sources for chemical detection. *Appl. Phys. Lett.*, Vol. 82, pp. 4648-4650.

[16] Jiang, Y., Jiang, W., Gu, L., Chen, X., and Chen, R. T. (2005) 80-micron interaction length silicon photonic crystal waveguide modulator. *Appl. Phys. Lett.*, Vol. 87, pp. 221105.

[17] Baba, T. (2008) Slow light in photonic crystals. *Nat. Photon.*, Vol. 2, pp. 465-473.

[18] Lodahl, P., Driel, A. F., Nikolaev, I. S., Irman, A., Overgaag, K., Vanmaekelbergh, D., and Vos, W. L. (2004) Controlling the dynamics of spontaneous emission from quantum dots by photonic crystals. *Nature*, Vol. 430, pp. 654-657.

[19] Yethiraj, A., and Blaaderen, A. V. (2003) A colloidal model system with an interaction tunable from hard sphere to soft and dipolar. *Nature*, Vol. 421, pp. 513-517.

[20] Fenollosa, R., and Meseguer, F. (2003) Non-close-packed artificial opals. *Adv. Mater.*, Vol. 15, pp. 1282-1285.

[21] Ho, K. M., Chan, C. T., and Soukoulis. C. M. (1990) Existence of a photonic gap in periodic dielectric structures. *Phys. Rev. Lett.*, Vol. 65, pp. 3152–3155.

[22] Yablonovitch, E., Gmitter, T. J., and Leung, K. M. (1991) Photonic band structure: the face-centered-cubic case employing nanospherical atoms. *Phys. Rev. Lett.*, Vol. 67, pp. 2295–2298.

[23] Sözüer, H. S., and Dowling. J. P. (1994) Photonic band calculations for woodpile structures. *J. Mod. Opt.*, Vol. 41, pp. 231–239.

[24] Zi, J., Xindi, Y., Yizhou, L., Xinhua, H., Chun, X., Xingjun, W., Xiaohan, L., and Rongtang, F. (2003) Coloration strategies in peacock feathers. *Proc. Nat. Acad. Sci. USA*, Vol. 100, pp. 12576-12578.

[25] Sanders, J. V. (1964) Colour of precious opal. *Nature*, Vol. 204, pp. 1151-1153.

[26] Huang, Y., Zhou, J., Su, B., Shi, L., Wang, J., Chen, S., Wang, L., Zi, J., Song, Y., and Jiang, L. (2013) Colloidal photonic crystals with narrow stopbands assembled from low-adhesive superhydrophobic substrates. *J. Am. Chem. Soc.*, Vol. 134, pp. 17053-17058.

[27] Liu, Y., Goebl, J., and Yin, Y. (2013) Templated synthesis of nanostructured materials. *Chem. Soc. Rev.*, Vol. 42, pp. 2610-2653.

[28] Wong, S., Kitaev, V., and Ozin, G. A. (2003) Colloidal crystal films: advances in universality and perfection. *J. Am. Chem. Soc.*, Vol. 125, pp. 15589-15598.

[29] Varghese, L. T., Xuan, Y., Niu, B., Fan, L., Bermel, P., and Qi, M. (2013) Enhanced photon management of thin-film silicon solar cells using inverse opal photonic crystals with 3d photonic bandgaps. *Adv. Opt. Mater.*, Vol. 1, pp. 692-698.

[30] Kumara, G. G., Senthilarasu, S., Lee, D. N., Kim, A. R., Kim, P., Nahm, K. S., Lee, S. H., and Elizabeth, R. N. (2008) Synthesis and characterization of aligned SiO_2 nanosphere arrays: Spray method. *Synthetic Metals*, Vol. 158, pp. 684-687.

[31] Arantes, T. M., Pinto, A. H., Leite, E. R., Longo, E., and Camargo, E. R. (2012) Synthesis and optimization of colloidal silica nanoparticles and their functionalization with methacrylic acid. *Colloids Surf. A*, Vol. 415, pp. 209-217.

[32] Lai, C. F., Lee, Y. C., Tsai, T. L., Chang, C. C., and Wu, M. K. (2014) Highly reliable and low-cost fabrication of warm-white LEDs using composite silica photonic crystals. *Int. J. Photoenergy*, Vol. 2014, pp. 846940.

[33] Lai, C. F., Lee, Y. C., and Tsai, T. L. (2015) Candlelight LEDs fabricated by using composite silica photonic crystals. *Optical Materials Express*, Vol. 5, pp. 307-313.

[34] Romanov, S. G., Bardosova, M., Povey, I. M., Pemble, M. E., and Torres, C. M. S. (2008) Understanding of transmission in the range of high-order photonic bands in thin opal film. *Appl. Phys. Lett.*, Vol. 92, 191106.

[35] Baryshev, A. V., Khanikaev, A. B., Fujikawa, R., Uchida, H., and Inoue, M. (2009) Diffraction processes in 3D photonic crystals based on thin opal films. *J. Mater. Sci Mater Electron*, Vol. 20, pp. S416-S420.

[36] Pavarini, E., Andreani, L. C., Soci, C., Galli, M., and Marabelli, F. (2005) Band structure and optical properties of opal photonic crystals. *Phys. Rev B*, Vol. 72, 045102.

[37] Romanov, S. G., Maka, T., Torres, C. M. S., Muller, M., Zentel, R., Cassagne, D., Martinez, J. M., and Jouanin, C. (2001) Diffraction of light from thin-film polymethylmethacrylate opaline photonic crystals. *Phys. Rev E*, Vol. 63, 056603.

[38] Lai, C. F., Chi, J. Y., Kuo, H. C., Yen, H. H., Lee, C. E., Chao, C. H., Yeh, W. Y., and Lu, T. C. (2009) Far-field and near-field distribution of GaN-based photonic crystal LEDs with guided mode extraction. IEEE J. Sel. Top. Quant. Electron., Vol. 15, pp. 1234-1241.

Chapter 4

Applications of Nano-Scale Plasmonic Structures in Design of Stub Filters — A Step Towards Realization of Plasmonic Switches

Hassan Kaatuzian and Ahmad Naseri Taheri

Additional information is available at the end of the chapter

http://dx.doi.org/10.5772/59877

1. Introduction

As electronic devices and circuits were shrinking to the nano-scale chips, some drawbacks hindered the reaching to the speed higher than tens of Giga Hertz, such as the higher power consumption, delay, and interference of the signals. However, the need for transmitting the huge amount of data over communication networks urged everyone in this area of technology to find an alternative for pure electronic devices. Despite rapidly developing the photonics technology, in practice the implementation of the photonic counterparts of the electronic devices encountered many problems such as inability to become integrated. The size of photonic devices halted on the half the wavelength of the operating signal due to the Diffraction limit of light.

In recent years, Surface Plasmon Polaritons (SPPs) have been considered as one of the most promising ways to overcome the diffraction limits of photonic devices. The field of manipulating the SPPs as the carrier of the signal is known as Plasmonics. In plasmonics, the advantages of electronics such as nano-scale component design are joint together with the benefits of photonics. Therefore, plasmonics shapes a key part of the fascinating field of nanophotonics, which discovers how electromagnetic fields can be confined over dimensions on the order of or smaller than the wavelength. It is constructed based on interaction of electromagnetic radiation and conduction electrons at metallic interfaces or in small metallic nanostructures, leading to enhance the confinement of optical field in a sub-wavelength dimension.

The strong interaction between microscopic metal particles and light has been utilized for thousands of years, the Lycurgus Roman Cup dating from 4'th century A.D. [1], brightly colored stained-glass window panels by annealing metallic salts in transparent glass [1]. The practical instructions for this ancient "nanofabrication" technique survive from as early as the

8th century A.D. [1]. From the beginning of the 20th century many phenomena related to the unusual optical properties of particles and spheres were explained; the blue color of the sky in terms of a simple derivation of the scattering power of spheres by Rayleigh [2], the bright color of metal glass by Garnett [3], presentation of a general formulation for the scattering of light from spherical surface by Gustav Mie [4]. In 1957, Ritchie [5] described surface plasmons in thin films in terms of electron energy loss spectroscopy. Otto [6], Kretschmann and Raether [7] explained the procedures for exciting surface plasmons on thin films optically in 1968. In 1974, the enhancement of Raman scattering from molecules influenced by the enhanced local fields at a rough metal surface was first observed by Fleischmann et al. [8].

In recent years, the applications of the surface plasmons are extended in optical integrated circuits and create a rapid developing field of photonics known as Plasmonics. Many researches are being carried out on developing plasmonic structures. Plasmonic switches and modulators are the most important components used for light routing and switching in the rapidly developing area of broadband optical communications and designing ultra-high speed switching devices, where sub-femtosecond interaction of light with matter is our main design concern [9]. Numerous plasmonic devices and components are introduced and huge of research are published. By exciting of the surface plasmons at the boundary of dielectric and metallic materials, an optical field can be confine into a subwavelength dimension [10]. A plasmonic circuit transforms the light into surface plasmons at the same frequency but reduced dimensions, in addition to guidance that is more flexible and highest speed.

In the next section of this chapter, Metal-Optics Electromagnetism is discussed. In section 3, wave analysis is used for explaining the propagation of surface plasmon polaritons. Section 4 is dedicated to investigate the various schemes of plasmonic guiding and their characteristics. In section 5, the modeling methods of plasmonic waveguides are introduced, and in section 6, a case study of plasmonic application in designing a switch is presented.

2. Metal-optics electromagnetism

Surface plasmons polaritons (SPPs) are surface electromagnetic modes that propagate at the interface of a dielectric with real electric permittivity ε_1 and a metal with permittivity $\varepsilon_2(\omega)<0$ as shown in figure 1.

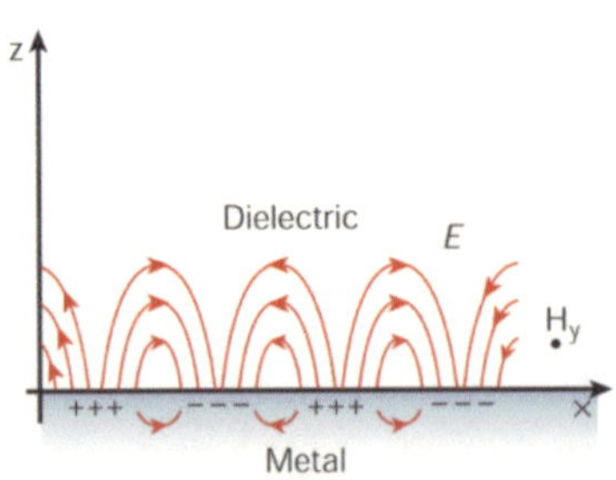

Figure 1. Electric and magnetic field distributions and charge oscillations at the dielectric/metal interface. SPPs propagate along the x-direction.

http://dx.doi.org/10.5772/59877

The complex permittivity or complex dielectric function can be expressed as [11]:

$$\varepsilon = \varepsilon' + i\varepsilon'' \tag{1}$$

where ε' is the real part and ε'' is th imaginary part of the permitivitty. In addition, the response of the material to the incoming optical field is expressed as the complex refractive index as [12]:

$$N = n + ik \tag{2}$$

where n and k are the real and imaginary part of the refractive index. The equations (1) and (2) are related to each other by:

$$\varepsilon' = n^2 - k^2 \tag{3}$$

$$\varepsilon'' = 2nk \tag{4}$$

These parameters are known as optical constants of the material however, in many of the materials they change with the frequency of the incident optical field. Especially in metals, the dielectric parameters are strongly dependent to the optical frequency. In order to use the permittivity of metals in the calculations of the spectrometry some mathematical models are introduced, such as Lorentz model, Lorentz-Drude model andExtended Drude model.

2.1. Lorentz model

In plasmonic, the optical properties of material can be expressed as a classical physical model of the microscopic structure [3], in which the charge carriers are assumed as damped harmonic oscillators subjected to incident electromagnetic field as driving forces. In the Lorentz model a charge carrier is supposed to have a mass of **m**, charge **e** and displacement from equilibrium **x**. The force on this particle as a linear spring force is $F = K.x$, the velocity dependent damping is $F = b.\dot{x}$ and the driving force by any incident light is $F = e\boldsymbol{E}$. Therfore, the equation of motion is [12, 13]:

$$m\ddot{x} + b\dot{x} + Kx = e\boldsymbol{E} \tag{5}$$

normalizing the above equation by mass:

$$\ddot{x} + \gamma + \omega_0^2 x = \frac{e}{m}\boldsymbol{E} \tag{6}$$

where $\omega_0^2 = K / m$ and $\gamma = b / m$. By solving the equation (6) as a time-harmonic and substitution of $\dot{x} \leftrightarrow -i\omega x$ and $\ddot{x} \leftrightarrow -\omega^2 x$:

$$x = \frac{\left(\frac{e}{m}\right)\boldsymbol{E}}{\omega_0^2 - \omega^2 - i\gamma\omega} \tag{7}$$

Supposing N particles per volume V and dipole $p = ex$ for each particle, the polarization is $\boldsymbol{P} = \left(\frac{N}{V}\right)p$. Therfore, equation (7) can be expressed:

$$\boldsymbol{P} = \frac{\omega_p^2}{\omega_0^2 - \omega^2 - i\gamma\omega}\varepsilon_0\boldsymbol{E} \tag{8}$$

where ω_p is the plasma frequency and is defined as:

$$\omega_p^2 = \frac{\left(\frac{N}{V}\right)e^2}{m\varepsilon_0} \tag{9}$$

From the Maxwell's equation, we have:

$$\varepsilon = 1 + \chi \tag{10}$$

$$\boldsymbol{P} = \varepsilon_0\chi\boldsymbol{E} \tag{11}$$

Therefore, from the equations (8), (10) and (11) the Lorentz model for the dielectric function is derived:

$$\varepsilon_{Lorentz}(\omega) = 1 + \frac{\omega_p^2}{\omega_0^2 - \omega^2 - i\gamma\omega} \tag{12}$$

2.2. Multi-oscillator Drude-Lorentz model

Here, the model is the superposition of j+1 individual oscillators [12]:

$$\varepsilon_{Drude-Lorentz}(\omega) = 1 - \frac{f_0\omega_{p,0}^2}{\omega^2 + i\gamma_0\omega} + \sum_{j=1}^{j_{max}} \frac{f_j\omega_{p,j}^2}{\omega_j^2 - \omega^2 - i\gamma_j\omega} \tag{13}$$

where f_j's are the oscillator strengths, $\omega_{p,j}$'s are the plasma frequencies, γ_j's are damping rates and ω_j's are the oscillator frequencies. The second term of equation (13) is the Drude term and represents the free electrons, which are not under the restoring force. Using 5 terms of the Lorentz model constructs an excellent approximation of optical properties of metals and many semiconductors in the visible spectrum [12].

3. Wave approach for the propagation of the surface plasmon polaritons

Surface plasmon polaritons are electromagnetic waves propagating at the interface between a dielectric and a metal evanescently bounded in the perpendicular direction [4]. These electromagnetic surface waves are created from the coupling of the optical fields to oscillations of the metal's electron plasma. Based on the dispersion relation and the spatial field distribution, the surface plasmons are described quantitatively.

3.1. Propagating the SPPs at a single metal-insulator interface

3.1.1. Deriving the dispersion relation

By solving the Maxwell's equation at a single metal-insulator interface, the wave equation is produced, in which for the transverse magnetic (TM or p) modes is [14]:

$$\frac{\partial^2 H_y}{\partial z^2} + \left(k_0^2 \varepsilon - \beta^2\right) H_y = 0 \tag{14}$$

and for transverse electric (TE or s) modes is:

$$\frac{\partial^2 E_y}{\partial z^2} + \left(k_0^2 \varepsilon - \beta^2\right) E_y = 0 \tag{15}$$

where $k_0 = \omega / c_0$ is the wave vector of the propagating wave in vacuum. Now, we should consider a simple flat boundary between an insulator (z>0) with a real positive permittivity ε_2 and a metal (z<0) with a complex permittivity $\varepsilon_1(\omega)$ (for metals $Re\{\varepsilon_1(\omega)\}<0$). Supposing the condition of propagating wave bounded to the interface with evanescent falloff in z-direction, the solutions for TM waves in $z > 0$ is [14]:

$$H_y(z) = A_2 e^{i\beta x} e^{-k_2 z} \tag{16}$$

$$E_x(z) = iA_2 \frac{1}{\omega \varepsilon_0 \varepsilon_2} k_2 e^{i\beta x} e^{-k_2 z} \tag{17}$$

$$E_z(z) = -A_1 \frac{\beta}{\omega\varepsilon_0\varepsilon_2} k_2 e^{i\beta x} e^{-k_2 z} \tag{18}$$

and for z < 0 is:

$$H_y(z) = A_1 e^{i\beta x} e^{k_1 z} \tag{19}$$

$$E_x(z) = -iA_1 \frac{1}{\omega\varepsilon_0\varepsilon_1} k_1 e^{i\beta x} e^{k_1 z} \tag{20}$$

$$E_z(z) = -A_1 \frac{\beta}{\omega\varepsilon_0\varepsilon_1} k_2 e^{i\beta x} e^{k_1 z} \tag{21}$$

where $k_i \equiv k_{z,i}$ is the component of the wave vector at z-direction. The reciprocal value of the k_i, $\hat{z} = 1/|k_z|$, is evanescent decay length of the fields in the perpendicular of the propagation [14]. The continuity condition of H_y and εE_z at the interface implies that $A_1 = A_2$ and

$$\frac{k_2}{k_1} = -\frac{\varepsilon_2}{\varepsilon_1} \tag{22}$$

Because at the interface of insulator and metal $Re\{\varepsilon_2\}>0$ and $Re\{\varepsilon_1\}<0$ the propagating waves confines at the surface. Based on equation (14) for TM waves at metal and insulator [14],

$$k_1 = \beta^2 - k_0^2\varepsilon_1 \tag{23}$$

$$k_2 = \beta^2 - k_0^2\varepsilon_2 \tag{24}$$

Therefore, by combining the equations (22), (23) and (24), the dispersion relation of the SPPs propagating at the metal-insulator yields

$$\beta = k_0\sqrt{\frac{\varepsilon_1\varepsilon_2}{\varepsilon_1 + \varepsilon_2}} \tag{25}$$

3.1.2. Forbidden modes for surface plasmons

Based on equation (15), the field components for TE modes are

$$E_y(z) = A_2 e^{i\beta x} e^{-k_2 z} \tag{26}$$

$$H_x(z) = -iA_2 \frac{1}{\omega\mu_0} k_2 e^{i\beta x} e^{-k_2 z} \tag{27}$$

$$H_z(z) = A_2 \frac{\beta}{\omega\mu_0} k_2 e^{i\beta x} e^{-k_2 z} \tag{28}$$

for z > 0, and

$$E_y(z) = A_1 e^{i\beta x} e^{k_1 z} \tag{29}$$

$$H_x(z) = iA_1 \frac{1}{\omega\mu_0} k_1 e^{i\beta x} e^{k_1 z} \tag{30}$$

$$H_z(z) = A_1 \frac{\beta}{\omega\mu_0} e^{i\beta x} e^{k_1 z} \tag{31}$$

In calculations of the equations (26) to (31), the following equations are used:

$$H_x = i\frac{1}{\omega\mu_0}\frac{\partial E_y}{\partial z} \tag{32}$$

$$H_z = \frac{\beta}{\omega\mu_0} E_y \tag{33}$$

From the continuity of E_y and H_x at the interface

$$A_1(k_1 + k_2) = 0 \tag{34}$$

and because of the confinement condition requires $Re\{k_1\}>0$ and $Re\{k_2\}>0$, hence $A_1=0$ and $A_2=0$; therefore, there are no TE modes at the surface. That means, the surface plasmon polaritons propagate only in TM polarization [14].

3.1.3. Propagation Length of SPPs

An important parameter in comparing plasmonic waveguide structures is the **propagation length** of a supported mode. As a surface plasmon propagates along the surface, it quickly loses its energy to the metal due to absorption. The intensity of the surface plasmon decays with the square of the electric field, so at a distance x, the intensity has decreased by a factor of $\exp(2Im(\beta_z)z)$ [15]. The propagation length is defined as the distance for the surface plasmon to decay by a factor of 1/e. This condition is satisfied at a length

$$L_{SPP} = \frac{1}{2Im(\beta_z)} \tag{35}$$

The typical value for propagation length in plasmonic metal-insulator waveguides is up to 100μm in visible regime [14].

4. Plasmonic waveguide schemes

The fundamental element in integrated plasmonic circuitry, as in photonics, is the waveguide. Because the necessity of existence of metal layer and various structural possibilities of using metal layer, many of schemes are developed in the recent years [11]. Each of the schemes has some benefits and drawbacks, in which make a trade-off between their characteristics. Two main characteristics of a waveguide are the confinement area of the mode and the propagation length. This features of the plasmonic waveguides are in opposite and there are a trade-off between them.

Figure 2. Optical field distribution at cross section of insulator-metal waveguide

In this section, some of the well-known types of plasmonic waveguides structures will be briefly introduced. In each case, the advantage(s) and weakness(s) of the waveguide are investigated.

4.1. Insulator-Metal (IM) waveguide

The simplest structure for guiding the plasmon polaritons is the Insulator-Metal (IM) waveguide (figure 2). This scheme consists a thin film metal coated on a simple insulator strip waveguide [11]. The silicon core of the waveguide has dimension about 300nm×300nm and the cladding of the waveguide is the air. The metal cap has a 50nm thickness and can be gold or silver. The structure placed on a SiO_2 substrate. The propagation length of this fundamental scheme is 2μm at 1550nm [11]. In this structure, a relative good balance exists between propagation length and confinement. In addition, the material used in fabrication allows feasibility of integrated plasmonic circuitry [11].

4.2. Dielectric-loaded SPP waveguide

The dielectric-loaded SPP waveguides (DLSPPWs) are formed by placing a polymethylmethacrylate (PMMA) ridge with a cross section of 500 nm in width by 600nm in height as the dielectric material on top of a 65nm thick and 3μm wide gold strip [16] (figure 3a). This scheme provides a full confinement in the plane perpendicular to the propagation direction. This geometry is one of the most popular for plasmonic waveguides because PMMA can work as both resist and the dielectric core for the DLSPPW [16]. There is a convenient and simple process to fabricate plasmonic devices based on DLSPPW using deep-ultraviolet (UV) lithography for those devices with larger dimensions working at telecom wavelengths or standard E-beam lithography (EBL) for those working in the near-infrared. In addition, due to the physical dimension and therefore the mode size increase, this structure has slightly better propagation length for SPP's going from 5μm to 25μm.

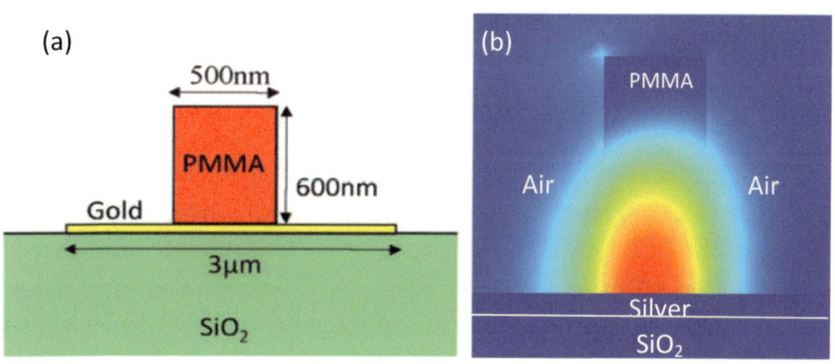

Figure 3. (a) Schematic representation of the DLSPPW cross section [16]. (b) Mode distribution of DLSPPW of optimal configuration.

Compared with other waveguide schemes, DLSPPWs are well-matched with different dielectrics and have a rather good trade-off between mode confinement and propagation distance [16]. These characteristics make them suitable for realization of dynamic components by utilizing of material (e.g., thermo- and electro-optic) effects, while strong mode confinement and long propagation distances are required for realization of compact and complex plasmonic circuits [16].

4.2.1. Long-Range Dielectric-Loaded Surface Plasmon Polaritons waveguides (LR-DLSPPs)

The long-range plasmonic waveguide is, unlike previous studies of DLSPPWs, based on a metal film with finite width [17], i.e., a metal strip, and can thus be considered a hybrid plasmonic waveguide (discussed later). The LR-DLSPPW configuration consists of a dielectric ridge placed on top of a thin metal strip, which is supported by a dielectric film (figure 4a). The entire structure is supported by a low-index substrate that ensures mode confinement to the dielectric ridge and underlying dielectric film [17].

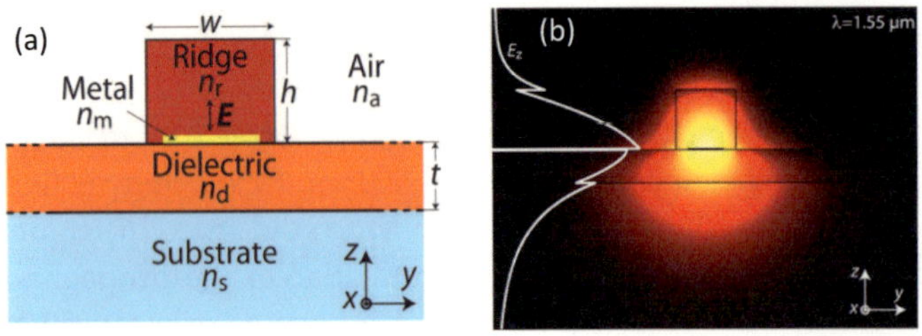

Figure 4. (a) Layout of the LR-DLSPPW structure, with a dielectric ridge on top of a thin metal stripe deposited on an underlying dielectric layer supported by a low-index glass substrate, (b) optical field distribution in the cross-section of the LR-DLSPPW [17].

Because of varying the thickness of the PMMA layer and hence, balancing the mode field on either side of the metal strip (figure 4b), this structure minimizes the losses and increase the propagation length to L = 3100μm, while retaining strong mode confinement [17].

4.3. Hybrid plasmonic waveguide

As a solution to the issue of propagation loss of SPPs, a new waveguide known as hybrid plasmonic waveguide (HPWG) is presented [18]. Figure 5a shows the cross section of a two dimensional HPWG. It consists of a high index region (silicon) disjointed from a silver surface by a low index layer (SiO_2). The close area of the silver-silica interface and the silicon slab results in coupling of the SPP mode and dielectric waveguide mode supported by these two structures. Figure 5b shows the resulting hybrid mode. The guide also supports a conventional TE mode, which is shown in figure 5c. The mode sizes for both TE and TM modes are comparable in this case and are very similar to mode size achievable in case of silicon waveguide. HPWG offers a number of benefits [18]: it has an improved compromise between loss and confinement compared to purely plasmonic waveguides, and is compatible with silicon on insulator technology. The power of the TE and TM modes in the HPWG are concentrated in two different layers, therefore their guiding characteristics can be controlled in different manners by altering the material properties, and waveguide dimensions of the layers [18].

http://dx.doi.org/10.5772/59877

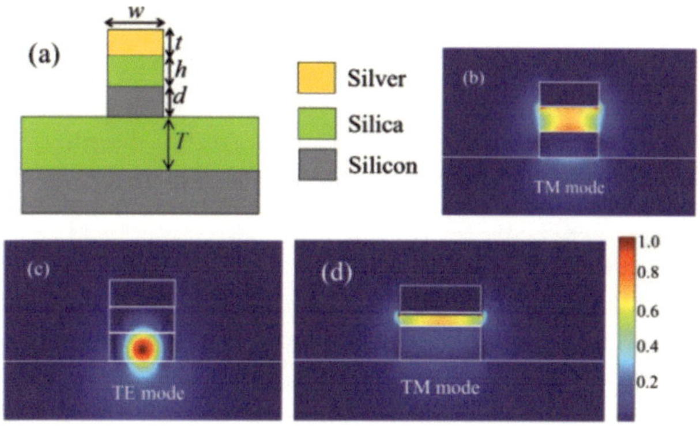

Figure 5. (a) Schematic of a hybrid waveguide. (b) and (c) power density profile for the TM and TE modes respectively for waveguide dimensions are w = 350 nm, t = 200 nm, h = 150nm, d = 150 nm and T = 2μm. (d) power density profile for the TM mode for w = 350 nm, t = 200 nm, h = 150 nm, d = 45 nm and T = 2μm. Wavelength of light is 1550 nm [18].

4.4. Metal-insulator-metal waveguide

In contrast to the hybrid waveguides, metal-insulator-metal (MIM) structure has a very good modal confinement in cost of a short propagation length. A MIM waveguide is a dielectric slot sandwiched as core between two layers of metal as cladding. The coupling of two SPP's from two metal-dielectric boundaries is the only aspect in common. In the MIM structure, these two SPP's coupled into the central dielectric slot and thus gives rise to a huge field concentration. However, due to the close proximity of the mode with both metal layers, the losses are extremely large. A typical value for propagation length is in the few microns as in the IM case. The most noticeable advantage of the MIM mode is its sub-wavelength size.

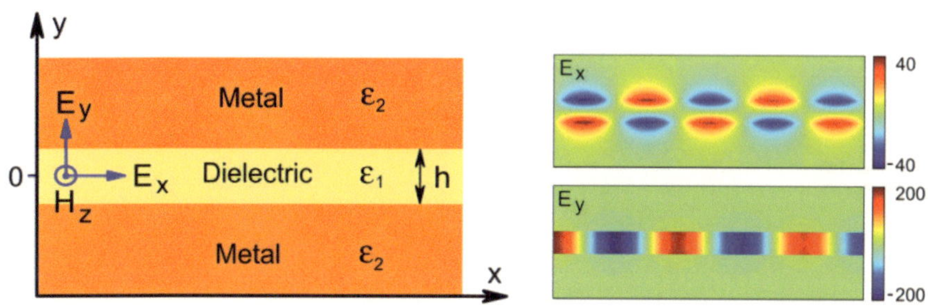

Figure 6. Left panel: Schematic of an MIM waveguide with dielectric layer of thickness h and permittivity ε_1 sandwiched between two metallic layers of permittivity ε_2. Right panel: Density plots of longitudinal (E_x) and transverse (E_y) electric fields corresponding to the fundamental anti-symmetric SPP mode [19].

In addition, in the MIM waveguides one could make any sharp geometries or bending regardless of wavelength of the incident optical field [20]. MIM subwavelength plasmonic waveguide bends and splitters have low loss over a wide frequency range [20]. While metals are naturally lossy, the bounded SP modes of a single insulator-metal interface can propagate over several microns under optical incident fields [21]. In such a geometry, the field skin depth increases exponentially with wavelength in the insulator but is almost constant (~25nm) in the metal for visible and near-infrared excitation frequencies. Not unlike conventional waveguides, these metal-insulator-metal (MIM) structures guide light via the refractive index differential between the core and cladding [21]. However, unlike dielectric slot waveguides, both plasmonic and conventional waveguiding modes can be accessed, depending on transverse core dimensions. MIM waveguides may thus allow optical mode volumes to be reduced to subwavelength scales—with minimal field decay out of the waveguide physical cross section— even for frequencies far from the plasmon resonance.

4.4.1. Dispersion relation for MIM waveguides

The modal analysis of planar multilayer structures can be solved via the vector wave equation under constraint of tangential **E** and normal **D** field continuity [21]. For unpolarized waves in a three-layer symmetric structure, the electromagnetic fields take the form [21]:

$$E(x,z,t) = \left(E_x\hat{x} + E_y\hat{y} + E_z\hat{z}\right)e^{i(k_x - \omega t)} \tag{36}$$

$$B(x,z,t) = \left(B_x\hat{x} + B_y\hat{y} + B_z\hat{z}\right)e^{i(k_x - \omega t)}, \tag{37}$$

with E_y, B_x and B_z identically zero for transverse magnetic (TM) polarization and E_x, E_z and B_y identically zero for transverse electric (TE) polarization. Inside the waveguide, the field components may be written as [21]:

$$\begin{gathered}
E_x^{in} = e^{-ik_{z1}z} \pm e^{ik_{z1}z}, \\
E_y^{in} = 0, \\
E_z^{in} = \left(\frac{k_x}{k_{z1}}\right)\left(e^{-ik_{z1}z} \mp e^{ik_{z1}z}\right), \\
B_x^{in} = 0, \\
B_y^{in} = \left(\frac{-\omega\varepsilon_1}{ck_x}\right)\left(e^{-ik_{z1}z} \mp e^{ik_{z1}z}\right), \\
B_z^{in} = 0,
\end{gathered} \tag{38}$$

for the TM polarization and as:

$$\begin{aligned}
E_x^{in} &= 0, \\
E_y^{in} &= e^{-ik_{z1}z} \pm e^{ik_{z1}z}, \\
B_z^{in} &= 0, \\
B_x^{in} &= \left(\frac{-k_{z1}c}{\omega}\right)\left(e^{-ik_{z1}z} \mp e^{ik_{z1}z}\right), \\
B_y^{in} &= 0, \\
B_z^{in} &= \left(\frac{k_x c}{\omega}\right)\left(e^{-ik_{z1}z} \pm e^{ik_{z1}z}\right),
\end{aligned} \tag{39}$$

for the TE polarization. Outside the waveguide, the components are given [21]:

$$\begin{aligned}
E_x^{out} &= \left(e^{-ik_{z1}d/2} \pm e^{ik_{z1}d/2}\right), \\
E_y^{out} &= 0, \\
E_z^{out} &= \left(\frac{\varepsilon_1 k_x}{\varepsilon_2 k_{z1}}\right)\left(e^{-ik_{z1}d/2} \mp e^{ik_{z1}d/2}\right)e^{ik_{z2}\left(z-\frac{d}{2}\right)}, \\
B_x^{in} &= 0, \\
B_y^{out} &= \left(\frac{-\omega\varepsilon_1}{ck_x}\right)\left(e^{-ik_{z1}d/2} \mp e^{ik_{z1}d/2}\right)e^{ik_{z2}\left(z-\frac{d}{2}\right)}, \\
B_z^{out} &= 0,
\end{aligned} \tag{40}$$

for TM polarization and as:

$$\begin{aligned}
E_x^{out} &= 0, \\
E_y^{out} &= \left(e^{-ik_{z1}d/2} \pm e^{ik_{z1}d/2}\right)e^{ik_{z2}\left(z-\frac{d}{2}\right)}, \\
E_z^{out} &= 0, \\
B_x^{out} &= \left(\frac{-k_{z1}c}{\omega}\right)\left(e^{-ik_{z1}d/2} \mp e^{ik_{z1}d/2}\right)e^{ik_{z2}\left(z-\frac{d}{2}\right)}, \\
B_y^{out} &= 0, \\
B_z^{out} &= \left(\frac{k_x c}{\omega}\right)\left(e^{-ik_{z1}d/2} \pm e^{ik_{z1}d/2}\right)e^{ik_{z2}\left(z-\frac{d}{2}\right)},
\end{aligned} \tag{41}$$

for TE polarization. The in-plane wave vector k_x is defined by the dispersion relations [21]:

$$L+:\begin{cases}\varepsilon_1 k_{z2}+\varepsilon_2 k_{z1}\tanh\left(\dfrac{-ik_{z1}d}{2}\right)=0\ TM\\ k_{z2}+k_{z1}\tanh\left(\dfrac{-ik_{z1}d}{2}\right)=0\ \ TE\end{cases} \tag{42}$$

$$L-:\begin{cases}\varepsilon_1 k_{z2}+\varepsilon_2 k_{z1}\coth\left(\dfrac{-ik_{z1}d}{2}\right)=0\ TM\\ k_{z2}+k_{z1}\coth\left(\dfrac{-ik_{z1}d}{2}\right)=0\ \ TE\end{cases} \tag{43}$$

where k_{z2} is the momentum conservation:

$$k_{z1,2}^2=\varepsilon_{1,2}\left(\frac{\omega}{c}\right)^2-k_x^2 \tag{44}$$

and $L+$ and $L-$ denote the antisymmetric and symmetric tangential electric field with respect to the waveguide medium, respectively.

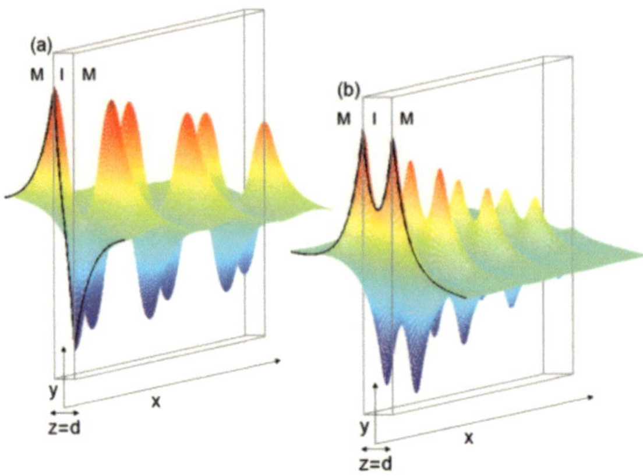

Figure 7. Geometry and characteristic tangential (x) electric field profiles for MIM slot waveguides; (a) Field antisymmetric mode, (b) Field symmetric mode [21].

In above analysis the structure is centered at z=0 with core thickness **d** and wave propagates along the positive x direction (figure 7). The core (cladding) is composed of material with complex dielectric constant ε_1 (ε_2). Since continuity of E_y forbids charge accumulation at the interface, TE surface plasmon waves do not generally exist in planar insulator-metal structures [21]. If SPPs are excited at an insulator-metal interface, electrons in the metal make a surface polarization that leads to a localized electric field. In insulator-metal-insulator (IMI) structures, electrons of the metallic core screen the charge configuration at each interface and maintain a near-zero (or minimal) field within the waveguide. Therefore, the surface polarizations on both side of the metal film persist in phase and there is no cutoff frequency for any transverse waveguide dimension [21]. In contrast, screening does not occur within the insulator core of MIM waveguides. At each insulator-metal interface, surface polarizations of each side of interface are independent, and therefore SPPs oscillations need not be energy- or wave-vector-matched to each other. As a result, for certain MIM dielectric core thicknesses, interface SPPs may not remain in phase but will exhibit a beating frequency; as transverse core dimensions are increased, "bands" of allowed energies or wave vectors and "gaps" of forbidden energies will be observed [21].

The effective refractive index n_{eff} of an MIM waveguide is complex. Its real part controls the guided wavelength λ_{MIM} and its imaginary part limits the propagation length L_{SPP} of SPPs through the relation [19]:

$$n_{eff} = \frac{\beta}{k} = \frac{\lambda}{\lambda_{MIM}} + i\frac{\lambda}{4\pi\lambda_{SPP}} \tag{45}$$

where $k = 2\pi / \lambda$, $\lambda = 2\pi c_0 / \omega$ and c_0 is the speed of light in vacuum. One can calculate β using the following dispersion relation for the TM-SPP modes [19]:

$$\tanh\left(\frac{ik_1 d}{2}\right) = \left(\frac{\varepsilon_2\, k_1}{\varepsilon_1\, k_2}\right)^{\pm 1} \tag{46}$$

where the signs ± correspond to symmetric and antisymmetric modes. Based on Equation (46), for $h \ll \lambda$, an MIM waveguide supports only a single antisymmetric mode that is similar to the fundamental TEM mode of a parallel-plate waveguide with perfect-electric-conductor (PEC) boundaries: The symmetric mode stops to exist because it experiences a cut-off in the reciprocal space [19]. Thus, in the deep subwavelength regime, an MIM waveguide operates as a single-mode plasmonic waveguide. Since this regime is the most interesting from the standpoint of nanophotonics applications, we assume in this chapter that the condition $h \ll \lambda$ is satisfied and focus on a single-mode plasmonic waveguide [19].

5. Modeling of plasmonic MIM waveguides

5.1. Challenges in modeling of plasmonics

Numerical simulation and modeling of plasmonic devices involves several challenges specific to plasmonics [22]. The permittivity of metals at optical wavelengths is complex, i.e. $\varepsilon(\omega) = \varepsilon'(\omega) + i\varepsilon''(\omega)$, and is a complicated function of frequency [22]. Thus, several simulation techniques which are limited to lossless, non-dispersive materials are not applicable to plasmonic devices [22]. In addition, the dispersion properties of metals have to be approximated by suitable analytical expressions, such as Drude model and Drude-Lorentz model. Furthermore, in surface plasmons propagating along the interface of a metal and an insulator, the field is confined at the interface, and decays exponentially perpendicular to the direction of propagation. Consequently, for numerical methods based on discretization of the fields, a very fine mesh resolution is required at the metal-dielectric interface [22]. Generally, simulation of plasmonic devices necessitates much finer grid resolution than modeling of low- or high-index-contrast dielectric devices, because of the high localization of the field at metal-dielectric interfaces of plasmonic components. The required grid size depends on the shape and feature size of the modeled plasmonic device, the metallic material used and the operating frequency.

5.2. General simulation methods for plasmonic components

Due to the wave nature of the surface plasmon polaritons, most of the simulation methods of plasmonic devices and circuits are the same as the radio frequency approaches; therefore, the most common methods in simulation and modeling of the plasmonics are Finite-Difference Time Domain (FDTD) [23], Finite Element Method (FEM) [24], Transmission Line Method (TLM) [19] and Coupled Mode Theory (CMT) [25]. Each of the methods has its advantages over a specific plasmonic structure, e.g. in order to simulation of a 3D hybrid structure of a ring resonator FDTD or FEM are more convenient and for modeling of a metal-insulator-metal stub filter the accurate and fast approach is TLM. Transient response of a structure may be modelled by simulating electromagnetic pulses in time domain, by methods such as FDTD. The discussion for investigation of the best way to simulate plasmonic waveguides is out of the scope of this chapter; however, some of the most widely used methods are briefly explained.

5.2.1. Finite-Difference Time Domain (FDTD)

Finite Difference Time Domain is a common technique in modelling electromagnetics problems [15]. It is considered easy to understand and implement [26]. It is a time-domain method, so depending on the excitation type used, it could cover a wide range of frequencies in a single run, and hence it is usually the method of choice for wideband systems [15]. This method was introduced first by Kane Yee. FDTD does not require the existence of Green's function and directly approximates the differential operators in Maxwell's equations on a grid discretized in time and space. FDTD is an explicit finite difference approach, i.e. no matrix equation is established, stored and solved. The field values at the next time step are given entirely in terms of the field at the current and the previous time steps [15].

FDTD discretizes into unit cells called "Yee cells" [27, 28]. In such Yee cells, Electric Fields are represented by the edges of a cube, where the faces of the cubic cell denote the magnetic field. Given the offset (in space) of the magnetic fields from the electric fields, the values of the field with respect to time are also offset. Time is distributed into small steps, which are corresponding to the amount of time needed for the fields to travel from one Yee cell to the next or less. FDTD method solves the Maxwell's equations using the relationship between the partial time and space derivatives [15]. Yee's algorithm solves both E and H in time and space using the Maxwell's curl equations [26]. The biggest advantage of FDTD is its simplicity of use and fast application.

http://dx.doi.org/10.5772/59877

5.2.2. Finite Element Method (FEM)

The finite element method (FEM) is a numerical technique, which approximates the solutions of differential equations [15]. In electromagnetics, FEM generally approaches the solution to the Maxwell's equations in the frequency domain, therefore is usually used with time-harmonic conditions. Furthermore, it is capable of time domain simulations. FEM can be derived using two methods [15]; First, the variational method, which finds a variational functional whose minimum/maximum/stationary point corresponds to the solution of the PDE subject to certain boundary conditions (a brief formulation is given in section 2.7.1). Second method is called the "weak formulation" in the literatures. It works by introducing a weighted residual error to one of the differentials in the PDE form of Maxwell's equations [29] and equating the sum of the error to zero. If the weighting functions are Dirac delta functions, the resulting procedure is similar to finite difference method. If the weighting functions are the basis functions, then the method is called the Galerkin's method [15].

Several commercial software are presented in which simulate the electromagnetic problems based on FEM, such as CST Microwave Studio, COMSOL Multiphysics and ANSYS.

5.2.3. Transmission Line Modelling (TLM) method

Transmission Line modelling is one of the time domain techniques that can solve the electromagnetic problems [15]. The concept of impedance and understanding the effects of waveguide discontinuities in terms of lumped circuit elements were crucial in this respect [22]. Although the properties of metals are quite different at optical wavelengths compared to the microwave, designs that are qualitatively similar to their low frequency counterparts have been demonstrated at optical frequencies.

In this method, a unit cell is formed by conceptually filling space with a network of transmission-lines in such a way that the voltage and current give information on the electric and magnetic fields [15]. A node represents the intersection point of the transmission-lines. At each time step, voltage pulses come to the node from each of the transmission-lines. These pulses are then scattered to produce a new set of pulses that become incident on adjacent nodes at the next time step. The relationship between the incident pulses and the scattered pulses is determined by the scattering matrix, which is set to be consistent with Maxwell's equations. Additional elements, such as transmission-line stubs, can be added to the node so that different material properties can be represented [15].

6. A case study: Electro-plasmonic switch based on MIM stub filter

6.1. Stub filters in MIM plasmonic structures

A stub is one of the key elements in microwave engineering and is employed in various microwave devices to reduce their size [30]. Some research groups have proposed a wavelength filter by using a stub structure in a photonic crystal waveguide [30, 31]. Such a structure may be employed in a plasmon waveguide to perform as a wavelength selective filter.

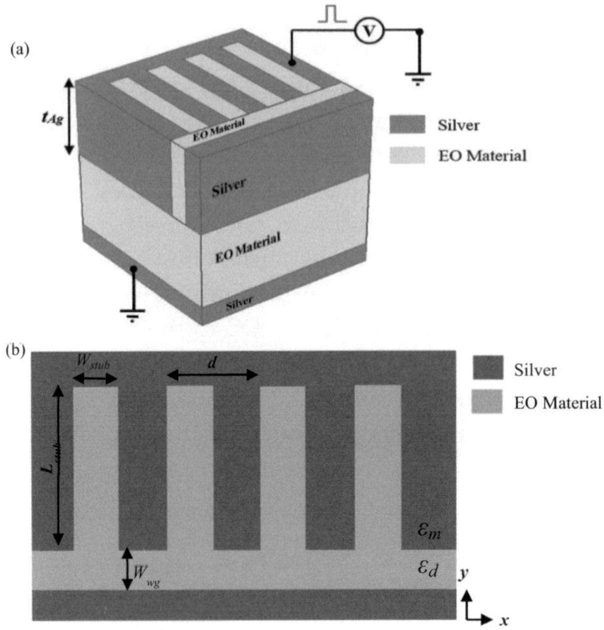

Figure 8. (a) three-dimension and (b) two-dimensional schematic depiction of the proposed Electro-Plasmonic switch [34].

Stub filters, also could be employed in order to make plasmonic switches [32-34]. MIM plasmonic waveguide structure helps to shrink the size of a stub filter to nanoscale and the stub structures cause to reduce the threshold voltage of the switch.

6.2. MIM Stub filter as a plasmonic switch

Figures 8(a) and 8(b) show the three and two-dimensional schematic of the proposed electro-plasmonic switch based on the MIM structure, respectively [34]. This switch is essentially an MIM stub filter, which comprises of a main waveguide and one or more stub(s) vertically connected to it. Several studies investigated the properties of the stub filters and many researches used their filtering characteristics [19, 35-36].

The width of the waveguide and stubs, the length of the stubs and the displacement between them, and the refractive index of the core adjust the properties of the stub filter [19]. Any change in these factors will change the transfer function of the filter; therefore, one can control the transmittance valley or peak of the filter with a change in above parameters. Thus, through altering the refractive index of the core by applying an electric field to an EO material one could have an EO switch. In an MIM structure, the metal cladding boosts the electro optic effect in the core; subsequently, it decreases the threshold voltage of the switch.

Here, it is supposed that the entire of the switch filled with an EO material known as 4-dimethyl-amino-Nmethyl-4-stilbazolium tosylate (DAST) as the core, with a linear refractive index of $n_d = 2.2$ and a large EO coefficient (dn/dE = 3.41 nm/V) [37].

Due to the low losses for the surface plasmons propagation, here silver is selected as the metal cladding of the waveguide [38]. In all simulations, the Drude-Lorentz model of the silver is utilized in order to obtain accurate results. A seven-pole Drude–Lorentz model is used in the wavelength range from 0.4μm to 2μm [19]:

$$\varepsilon_m(\omega) = 1 - \frac{\omega_p^2}{\omega(\omega + i\gamma)} + \sum_{n=1}^{5} \frac{f_n \omega_n^2}{\omega_n^2 - \omega^2 - i\omega\gamma_n} \tag{47}$$

where, ω_p=2002.6 *THz* is the bulk plasma frequency of silver and γ=11.61 *THz* is a damping constant [19]. Table 1 listed the other parameters. In addition, we used an analytical expression to calculate of the effective index of the fundamental TM_0 mode in the entire of the device [39]:

$$n_{eff} = \sqrt{\varepsilon_d}\left(1 + \frac{\lambda}{\pi w \sqrt{-\varepsilon_m}} \sqrt{1 + \frac{\varepsilon_d}{-\varepsilon_m}}\right)^{\frac{1}{2}} \tag{48}$$

where, w is the width of the waveguide and $\varepsilon_d = n_d{}^2$. The real and imaginary parts of the effective index for w =50nm over the wavelength range from 0.4 to 2μm is depicted in figure 9.

n	ω_n (THz)	γ_n (THz)	f_n
1	197.3	939.62	7.9247
2	1083.5	109.29	0.5013
3	1979.1	15.71	0.0133
4	4392.5	221.49	0.8266
5	9812.1	584.91	1.1133

Table 1. Parameters of the Drude–Lorentz Model for Silver [19].

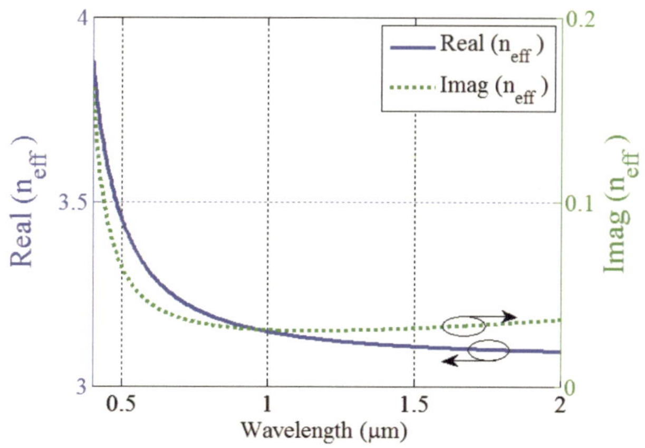

Figure 9. The real and imaginary part of the effective refractive index for MIM waveguide with the linear refractive index n_d =2.2 and W_{wg} = 50nm [34].

6.2.1. Modeling of the device using TLM

The Transmission Line Method (TLM) helps to study the operation of the MIM stub filters more accurately and faster [35], in comparison to other methods. In this method, we convert the plasmonic MIM structure in figure 10a to an equivalent circuit diagram (see figure 10b). This equivalent circuit is formed by a parallel connection of the characteristic impedance of an infinite MIM waveguides (Z_{MIM}) and the characteristic impedance of a finite and L_{spp} is the propagation length of the SPPs (equation (35)). The length of the stubs is the key point of the design of the stub filter and the electro-plasmonic switch [34]. Knowing that small increase in the real part of the refractive index leads to a red-shift in the transmittance spectrum of the filter, the transmission of the filter at wavelength λ=1550nm must be on the edge of a change, a falling edge of the magnitude for normally ON and a rising edge for normally OFF switch. Figure 11 shows the transmission of a typical 4-stub filter for different length of the stubs. The transmittance is defined as $T=P_{out}/P_{in}$ and is calculated using Transmission Line Method [35]. In the computation of the spectrum, the width of the main waveguide and the width of the stubs has the same value $W_{wg}= W_{stub}$=50nm, the distances between stubs is d=100nm and the thickness of the silver layer (or the depth of the structure) is t_{Ag}=200nm.

MIM waveguide (Z_S) terminated by Z_L accounts for reflection of the SPPs from the stub [19]:

$$Z_S = Z_{MIM} = \frac{\beta(w)w}{\omega\varepsilon_d\varepsilon_m} \tag{49}$$

$$Z_L = \sqrt{\frac{\varepsilon_m}{\varepsilon_d}} \times Z_S \tag{50}$$

$\beta(w)$ is the complex-valued propagation constant and describes the properties of the MIM waveguide. The circuit diagram in figure 10b may be modified to an equivalent form as shown in figure 10c by replacing Z_{MIM} and Z_S by an effective impedance (Z_{stub}). The value of Z_{stub} can be found from transmission-line theory and is given by [19]:

$$Z_{stub} = Z_S \frac{Z_L - iZ_S \tan\left(\beta(w)L_{stub}\right)}{Z_S - iZ_L \tan\left(\beta(w)L_{stub}\right)} \tag{51}$$

Now, we calculate the transmission spectra of an MIM stub filter with *N* identical stubs set apart by a distance *d*, the length of the stubs L_{stub}, the width of waveguide and stubs W_{wg} = W_{stub} = w [39]:

$$T = \left(P_+^{N-1}Q_+ - P_-^{N-1}Q_-\right)^{-2} \exp\left(-\frac{L}{L_{spp}}\right) \tag{52}$$

where

$$P_{\pm} = \frac{1}{2}\left[1 + \frac{Z_{MIM}}{2Z_{stub}} + \left(1 - \frac{Z_{MIM}}{2Z_{stub}}\right)\exp(2i\beta d) \pm R\right],$$

$$Q_{\pm} = \frac{1}{2R}\left(\left(1 + \frac{Z_{MIM}}{2Z_{stub}}\right)^2 - \left[1 + \left(\frac{Z_{MIM}}{2Z_{stub}}\right)^2\right]\exp(2i\beta d)\right) \pm \frac{1}{2}\left(1 + \frac{Z_{MIM}}{2Z_{stub}}\right)$$

$$R = \left\{\left[1 + \frac{Z_{MIM}}{2Z_{stub}} + \left(1 - \frac{Z_{MIM}}{2Z_{stub}}\right)\exp(2i\beta d)\right]^2 - 4\exp(2i\beta d)\right\}^{\frac{1}{2}},$$

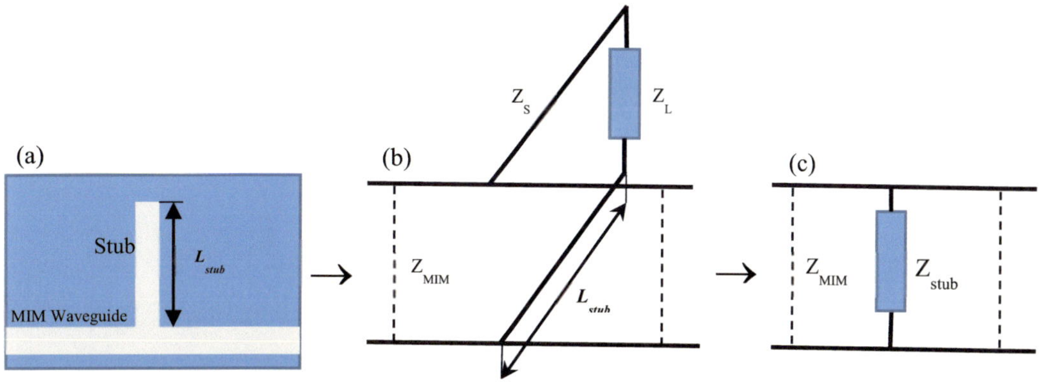

Figure 10. The diagram of an MIM waveguide with a single stub coupled perpendicularly to the waveguide axis. (b) The corresponding transmission-line representation. (c) The simplified circuit model. Z_{MIM} and Z_S correspond to the characteristic impedance of the MIM waveguide and the stub, respectively, Z_L accounts for reflection of the SPPs from the stub end and Z_{stub} is the effective stub impedance [19].

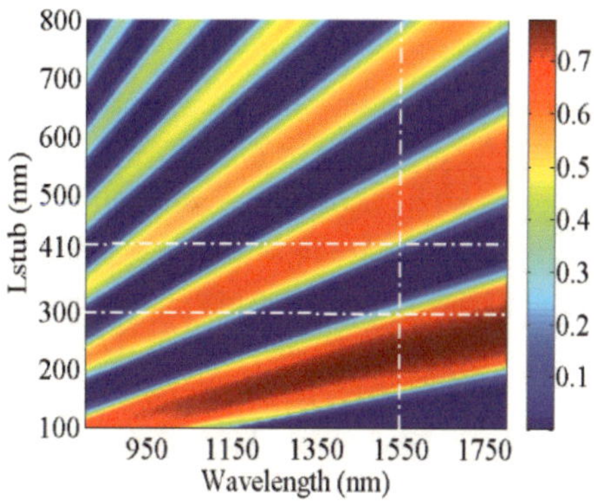

Figure 11. The transmittance spectra of the 4-stub filter for different length of the stubs. The crossing points of the dashed-lines show the transmittance values for L_{stub}=300nm and L_{stub}=410nm when λ = 1550nm. By a red-shift in the spectrum, the magnitude of the transmittance rises at the point a while falls at the point b [34].

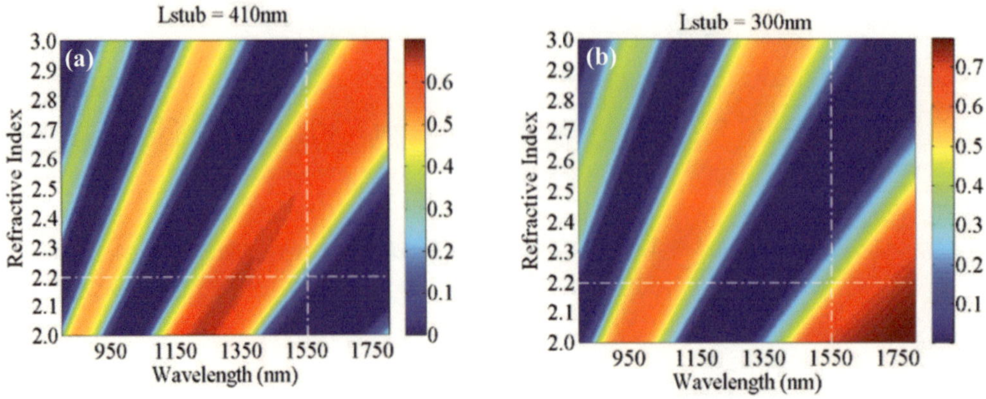

Figure 12. The transmittance spectra of the 4-stub filter for different refractive index with (a) L_{stub}= 410nm, and (b) L_{stub}= 300nm; the crossing of the dashed-lines indicate the transmittance of the filter with the correspond L_{stub}. Increment in the refractive index at 1550nm leads to (a) rise, (b) fall the magnitude of the transmittance [34].

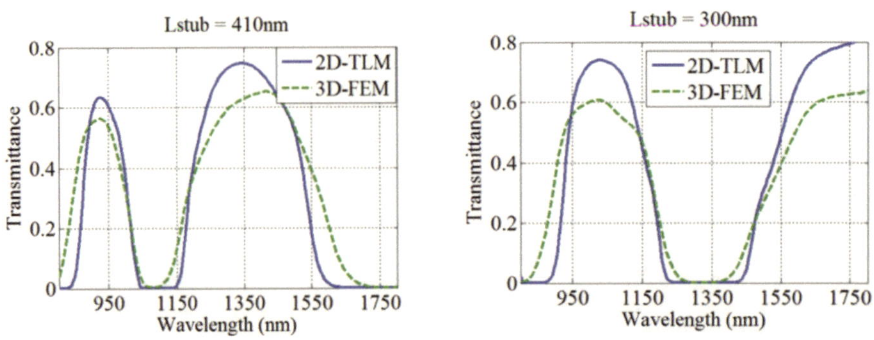

Figure 13. The transmittance spectra of the 4-stub filter with (a) L_{stub}= 410nm, and (b) L_{stub}= 300nm; in the calculation of the spectrums, 2D-TLM and 3D-FEM methods are used which show an acceptable conformity [34].

According to figure 11, by selecting the stub L_{stub}=300nm (point b) or L_{stub}=410nm (point a) at the communication wavelength λ=1550nm, the normal state of the switch would be ON or OFF, respectively. Slightly increment in the refractive index turns the switch OFF or ON, depending on the length of stubs (L_{stub}=300nm or L_{stub}=410nm, respectively). Figures 12a and 12b show the spectrum of the switch versus the refractive index for L_{stub}=300nm and L_{stub}=410nm, respectively. Figures 13a and 13b show the transmission spectrum for L_{stub}= 410nm and L_{stub}=300nm and shows reasonable accuracy of three-dimensional FEM simulations [34]. These figures admit the red-shift effect on the ON/OFF status of the switch. As shown in figure 8, the silver cladding is connected to the voltage V and therefore the electric field surrounds the main waveguide and all the stubs. The relation between applied voltages and the refractive index of EO material is expressed using equation [37]:

$$n = n_0 + \frac{dn}{dE}\left(\frac{V}{w}\right) \tag{53}$$

where n_0 is the linear refractive index of the EO material that equals 2.2, V is the applied voltage, and w is the width of waveguides and stubs [34]. The applied electric field alters the linear refractive index of the switch and as a result causes the red-shift. Figures 14a-d demonstrate the distribution of the optical field in the structure of the electro-plasmonic switch. In figures 14a and 14c the applied voltage is 0V and corresponds to the normal-state of both switches, while figures 14b and 14d show the electro-optic switch when the voltage is applied to the structure. The applied electric field changes the refractive index of the EO material; therefore, the optical field reflected back to the input port. Figure 15 depicts the transmittance of the normally ON and normally OFF switch as a function of the applied voltage at the wavelength 1550nm. As can be seen, the threshold voltage for switching is 10V [34].

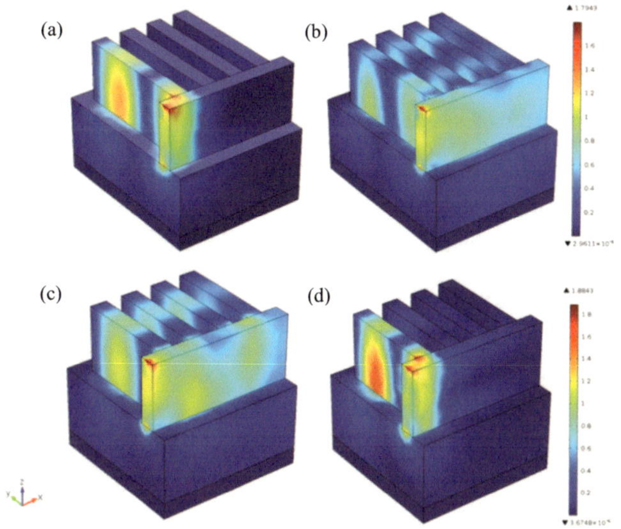

Figure 14. The distribution of the optical field in the structure of the electro-optic switch for L_{stub}=410nm when (a) V = 0V (b) V = 10V and L_{stub}=300nm when (c) V = 0V (d) V = 10V [34].

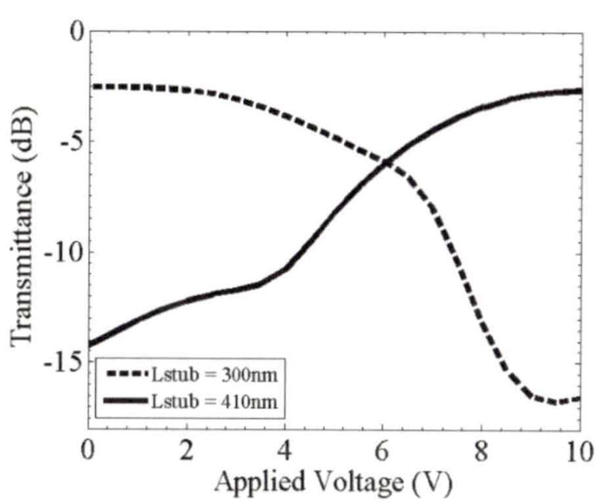

Figure 15. The transmittance of the proposed electro-optic switch versus the applied voltage at the telecom wavelength 1550nm; the solid and dashed line corresponds to EO switch with the L_{stub}=410nm (Normally OFF) and the L_{stub}=300nm (Normally ON), respectively [34].

Author details

Hassan Kaatuzian* and Ahmad Naseri Taheri

*Address all correspondence to: hsnkato@aut.ac.ir

Photonics Research Lab., Amirkabir university of Technology, Iran

References

[1] Atwater Harry, The promise of Plasmonics, Scientific American, April 2007.

[2] L. Rayleigh. Phil. Mag., 47, 1899.

[3] J. C. Maxwell Garnett. Colours in metal glasses and in metallic films. Philosophical Transactions of the Royal Society of London, 203, 1904.

[4] G. Mie. Beitrage zur optik truber medien, speziell kolloidaler metall°osungen. Annalen Der Physik, 25(3):377–445, 1908.

[5] R. H. Ritchie. Plasma losses by fast electrons in thin films. Physical Review, 106, 1957.

[6] A. Otto. Excitation of nonradiative surface plasma waves in silver by the method of frustrated total reflection. Zeitschrift fur Physik, 1968.

[7] E. Kretschmann and H. Raether. Radiative decay of non-radiative surface plasmons excited by light. Zeitschrift fur Naturforschung, 23A, 1968.

[8] M Fleischmann, P.J. Hendra, and A.J. McQuillan. Raman spectra of pyridine adsorbed at a silver electrode. Chemical Physics Letters, 26, 1974.

[9] Kaatuzian Hassan, "Quantum Photonics, A Theory for Attosecond Optics", AKU Press, ISBN: 978-964-463-273-0, 2012.

[10] D. K. Gramotnev and S. I. Bozhevolnyi, "Plasmonics beyond the diffraction limit," Nat.Photon. 4, 83-91 (2010).

[11] Zia, R., Schuller, J. A., Chandran, A., Brongersma, M. L.: Plasmonics: the next chip-scale technology, Materials Today 9, 20 (2006).

[12] Sweatlock L.A., Plasmonics: Numerical Methods and Device Applications. PhD Thesis. California Institute of Technology. 2008.

[13] Kaatuzian Hassan, "Photonics, vol-1", AKU Press, ISBN: 964-463-271-0, 2008.

[14] Maier S. A. Plasmonics: Fundamentals And Applications. Springer. 2007.

[15] Ahmet Arca. The design and optimisation of nanophotonic devices using the Finite Element Method. PhD Thesis. University of Nottingham. 2010.

[16] A. Kumar, J. Gosciniak, V. S. Volkov, S. Papaioannou, D. Kalavrouziotis, K. Vyrsokinos, J.-C. Weeber, et al., Dielectric-loaded plasmonic waveguide components: going practical, Laser Photon. Rev. 7 (6), pp. 938–951, 2013. doi:10.1002/lpor.201200113

[17] T. Holmgaard, J. Gosciniak, and S. I. Bozhevolnyi, "Long-range dielectric-loaded surface plasmon-polariton waveguides," Opt. Express 18(22), 23009–23015 (2010).

[18] Muhammad Zulfiker Alam, Hybrid Plasmonic Waveguides: Theory and Applications. PhD Thesis. University of Toronto. 2012.

[19] Pannipitiya, A., Rukhlenko, I. D., Premaratne, M., Hattori, H. T., Agrawal, G. P.: Improved transmission model for metal-dielectric-metal plasmonic waveguides with stub structure, Opt. Express, Vol. 18, No. 6, 6191 (2010).

[20] G. Veronis, and S. H. Fan, "Bends and splitters in metal-dielectric-metal subwavelength plasmonic waveguides," Appl. Phys. Lett. 87(13), 131102 (2005).

[21] J. A. Dionne, L. A. Sweatlock, and H. A. Atwater. Plasmon slot waveguides: Towards chip-scale propagation with subwavelength-scale localization. Physical Review B 73, 035407, 2006.

[22] G. Veronis, S. E. Kocaba ş, D. A. B. Miller, and S. Fan, " Modeling of plasmonic waveguide components and networks, " J. Comput. Theor. Nanosci. 6, 1808 – 1826 (2009).

[23] Y. Zhao, and Y. Hao, "Finite-difference time-domain study of guided modes in nanoplasmonic waveguides,"IEEE Trans. Antennas Propag. 55, 3070-3077 (2007).

[24] J. M. McMahon, A.-I. Henry, K. L. Wustholz, M. J. Natan, R. G. Freeman, R. P. Van Duyne, and G. C. Schatz, "Gold nanoparticle dimer plasmonics: finite element method calculations of the electromagnetic enhancement to surface-enhanced Raman spectroscopy," Anal. Bioanal. Chem. 394(7), 1819–1825 (2009).

[25] W.-P. Huang, "Coupled-mode theory for optical waveguides: an overview," J. Opt. Soc. Am. A 11, 963-983 (1994).

[26] A. Taflove and S. C. Hagness. Computational Electrodynamics: the finite-difference time domain method. Artech House, Norwood MA, 2000.

[27] Yee Kane S., "Numerical Solution of Initial Boundary Value Problems Involving Maxwells Equations in Isotropic Media," IEEE T. Antenn. Propag., vol. 14, 1966.

[28] A. Taflove, "Application of finite-difference time-domain method to sinusoidal steady-state electromagnetic-penetration porblems," IEEE Transactions on electromagnetic compatibility, vol. EMC-22, no. 3, 1980.

[29] D. B. Davidson, Computational Electromagnetics for RF and Microwave Engineering. Cambridge University Press, 2005.

[30] Y. Matsuzaki, T. Okamoto, M. Haraguchi, M. Fukui, and M. Nakagaki, "Characteristics of gap plasmon waveguide with stub structures," Opt. Express 16, 16314-16325 (2008).

[31] K. Ogusu and K. Takayama, "Transmission characteristics of photonic crystal waveguides with stubs and their application to optical filters," Opt. Lett. 32, 2185–2187 (2007).

[32] Min, C., Veronis, G.: Absorption switches in metal-dielectric-metal plasmonic waveguides, Opt. Express, Vol 17, No. 13, 10757 (2009).

[33] Naseri Taheri A. and Kaatuzian H. Design and simulation of a nanoscale electro-plasmonic 1 × 2 switch based on asymmetric metal – insulator – metal stub filters, Vol. 53, No. 27, Applied Optics, 2014.

[34] Naseri Taheri A. and Kaatuzian H. Numerical Investigation of a Nano-Scale Electro-Plasmonic Switch Based on Metal-Insulator-Metal Stub Filter. Opt Quant Electron. Springer 2014.

[35] Liu, J., Fang, G., Zhao, G., Zhang, Y., Liu S.: Surface plasmon reflector based on serial stub structure, Opt. Express, 17, 20134 (2009).

[36] Park, J., Kim, H., Lee, B.: High order plasmonic Bragg reflection in the metal-insulator-metal waveguide Bragg grating, Opt. Express, Vol. 16, No. 1, 413 (2008).

[37] Jin, Z. Y., Guange, H. X., Xian, M.: A Surface Plasmon Polariton Electro-Optic Switch Based on a Metal-Insulator-Metal Structure with a Strip Waveguide and Two Side-Coupled Cavities, Chin. Phys. Lett., Vol. 29, No. 6, 064214, 2012.

[38] Nielsen, M., Elezzabi, A.: Frequency-selective 3-D integration of nanoplasmonic circuits on a Si platform, Opt. Express, Vol. 20, No. 8, 8592 (2012).

[39] Collin, S., Pardo, F., Pelouard, J.: Waveguiding in nanoscale metallic apertures, Opt. Express, Vol. 15, No. 7, 4310, (2007).

Chapter 5

Equal Frequency Surface

G. Alagappan

Additional information is available at the end of the chapter

http://dx.doi.org/10.5772/59887

1. Introduction

A photonic crystal (PC) forbids propagation of light in a spectral range called photonic band gap. Within this frequency gap, PC behaves like an insulator for light. The insulating properties of the PC have been widely used in the developments of waveguides, fibers and cavities. For frequencies outside the photonic band gaps though light can propagate, the propagation properties (i.e. conducting properties) are uniquely different. The distinct conducting properties of the PC have led to discoveries of novel functionalities such as superprism effects, large angle polarization splitting, negative refraction, and superlensing.

The conducting properties of a PC can be best – analyzed using an equal frequency surface (EFS). The EFS is a surface in a three-dimensional (3D) PC, and a contour in a 2D PC. The gradient of the EFS plays a key role in determining the group velocity direction and hence, the propagation direction of light in the PC. Important developments on the EFS analysis for PCs can be found in [1-7].

EFS is typically obtained using a plane wave expansion methodology [8-9]. If the dielectric contrast of the PC is large, then a large number of plane waves is required to obtain EFSs with good accuracy. However, the distinct conducting properties, like superprism and beam splitting are normally well-pronounced in the PCs with a small dielectric modulation. [10-13]. For such PCs with small dielectric modulation, the requirement on the large number of plane waves to obtain the EFS can be relaxed.

In this chapter we will outline the theory of the EFS construction, and their applications in determining the conducting properties. For the sake of an easy understanding, throughout this chapter, we will use 2D PCs for illustrations and discussions. The chapter is organized as follows. The second section of the chapter focuses on the exact and approximate constructions of EFSs. Specifically, using one and two plane waves approximations [5], simple and handy analytical expressions for EFSs will be shown. It is worth mentioning that in the electronic energy band theory, the one plane wave technique is used to approximate Fermi surfaces of a metal, and it is known in the name of Harrison's principles [14-15]. In the third section, we will outline a graphical technique that uses EFSs to determine light propagation directions in the PC. Fourth section of the chapter elaborates on the applications of the EFSs. Particularly, we show the usage of the EFSs to describe superprism, effective negative refractive index mediums, negative refractions, and superlens phenomena in PCs. Finally, fifth section of the chapter gives a summary of the whole chapter.

2. EFS construction

In order to construct an EFS for a specific frequency, firstly, a photonic band structure has to be calculated. The photonic band structure can be calculated using a plane wave expansion method. For the details of this method, see [8-9]. The photonic band structure should contain frequencies for a dense number wavevectors in the first Brillouin zone (BZ). For a 2D PC, the EFS can be obtained by projecting the 2D photonic band structure onto the wavevector plane (k_x,k_y), keeping a constant frequency.

In a homogenous dielectric medium of refractive index n, the dispersion relation is $\omega=ck/n$, where ω, c, and k are the angular frequency, speed of light, and wavevector, respectively. In 2D, the dispersion relation is = $\omega^2=c^2(k_x^2+k_y^2)/n^2$. Thus, as shown in Figure 1, the EFSs of a homogeneous dielectric medium are circles. In a PC, the dispersion relation (i.e., the photonic band structure) is complicated, and therefore EFSs have to be calculated numerically.

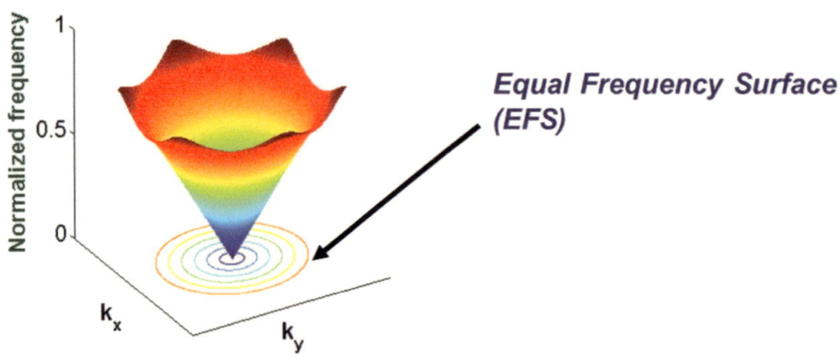

Figure 1. photonic band structure, and EFSs of a homogenous dielectric medium

Let's take examples of EFSs in a 2D square lattice PC made of circular silicon rods (dielectric constant=12.1) in an air matrix (dielectric constant=1.0). In such a system, the two polarizations of light can be decoupled. The band structure of the E – polarization (electric field perpendicular to the 2D periodic plane) is shown in Figure 2, for a silicon rod radius of 0.2. The vertical axis in this figure represents the normalized frequency, $\omega=a/\lambda$, with a and λ being the period and free space wavelength, respectively. As we can see from the 2D band structure, in the long wavelength limit (i.e., $a \ll \lambda$), the dispersion relationship is similar to a homogenous medium (Fig. 1).

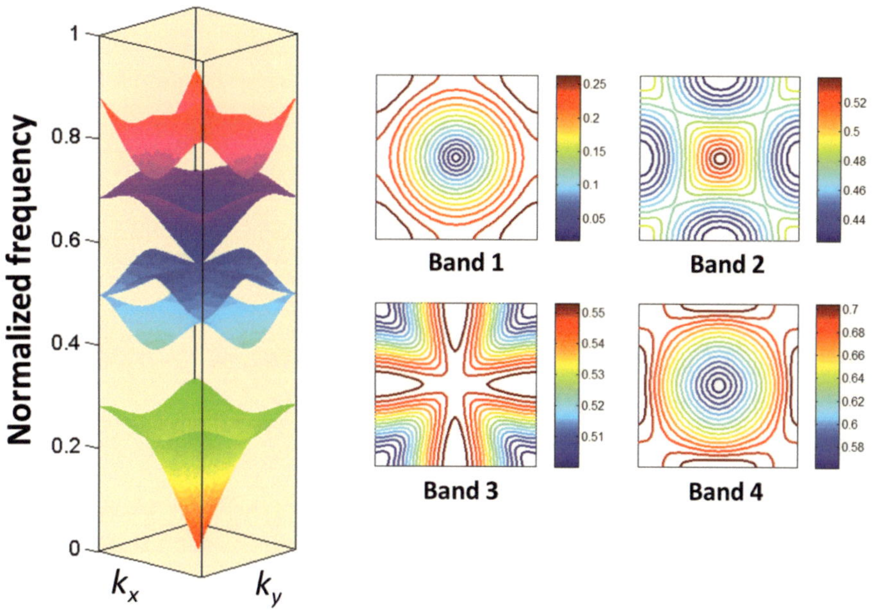

Figure 2. Photonic band structure in 2D and EFSs (in 2D these are constant frequency contours) of a 2D PC [circular silicon rods with a radius of 0.2, air matrix]. The color bars indicate the normalized frequencies

2.1. Approximation techniques

In a 3D PC made of either isotropic or anisotropic materials, it is known that the Maxwell equations corresponding to two independent polarizations of light are coupled [9,16-17]. For a 2D PC made of isotropic materials, we can always decouple the equations into two independent equations, corresponding to the two independent polarizations [9, 16]. On the other hand, for a 2D PC made of an anisotropic material, such a decoupling of polarizations is not possible in general. However, by restricting one of the principal axes of the anisotropic material to be perpendicular to the periodic plane of the 2D PC and the other two principal axes residing in the periodic plane, decoupling is possible [6,18-20]. Detail mathematical treatment of the polarization decoupling in a 2D PC made of anisotropic materials can be found in [6].

Assuming the polarizations can be decoupled, the photonic band structure of the *H*-polarization (magnetic field perpendicular to the periodic plane) or the *E*-polarization can be found by solving the differential-equation in the form [9,6,16],

$$\hat{D}[\mathbf{H}(\mathbf{r})] = (2\pi\omega / a)^2 [\mathbf{H}(\mathbf{r})], \tag{1}$$

for a given wavevector in the first BZ. In Eqn. 1, $\mathbf{H}(\mathbf{r})$ is the position ($\mathbf{r}$) dependent magnetic field vector, and $\hat{D}$ is a differential operator that depends on the polarization, and the dielectric constant profile of the PC. Using the plane wave expansion method [9,6-7,16], we can transform Eqn. 1 to a matrix eigen-value problem. Consequently, the frequency containing term $(2\pi\omega/a)^2=\Omega$ in Eqn. 1, can be written as a matrix vector product,

$$\Omega = \mathbf{h} \cdot \hat{M}\mathbf{h}. \tag{2}$$

In this equation, $\mathbf{h}=[h_1\ h_2...]'$, where h_n is the Fourier expansion coefficient of $\mathbf{H}(\mathbf{r})$ associated with the reciprocal lattice vector $\mathbf{G}_n$. The matrix element of $\hat{M}$ [6] is defined as

$$M_{mn} = \begin{cases} \langle k_m | \tilde{\beta}_r(m-n) | k_n \rangle, & H-\text{polarization} \\ |k_m||k_n| \beta(m-n), & E-\text{polarization} \end{cases}. \tag{3}$$

Here $|k_n\rangle = \mathbf{k} - \mathbf{G}_n$, with $\mathbf{k}=[k_x, k_y]'$. In Eqn. 3, <...> represents the inner product of the vector, matrix and vector. For the $H-$polarization, $\beta_r(n)$ is the 2 by 2 matrix defining the inverse Fourier transform coefficient of the tensor dielectric function [associated with $\mathbf{G}_n$], and similarly for the E-polarization, $\beta(n)$ is the inverse Fourier transform coefficient of the scalar dielectric function.

Eqn. 2 defines an EFS for a given frequency, and has to be evaluated using the entire basis of the reciprocal space, which is infinite in number. In practice, the number of basis (i.e. the number of plane waves) is limited to a number for which the corresponding result achieves a required degree of accuracy. In general, larger the dielectric modulation, larger the number of required plane waves. For PCs with weak dielectric modulations, the number of plane waves can be significantly less. In the following sections, we will use one and two plane waves to approximate the EFS of the weakly modulated PCs, and as we shall see, such approximations lead to handy analytical expressions.

2.1.1. *One plane wave approximation*

Firstly, we will elaborate the one plane wave approximation [5] for the H – polarization. Consider a 2D PC made of two materials with permittivity tensors $\tilde{\varepsilon}_a$ and $\tilde{\varepsilon}_b$, and assume the fill factor of the material with the permittivity tensor $\tilde{\varepsilon}_a$ is f. The one plane wave technique assumes a very weak dielectric modulation such that $\beta_r(m - n)$, can be approximated with only one plane wave (i.e., one Fourier component) as $\beta_r(m - n) = \delta_{mn}\tilde{\varepsilon}_0^{-1}$, where $\tilde{\varepsilon}_0$ is the averaged dielectric tensor,

$$\tilde{\varepsilon}_0 = \tilde{\varepsilon}_a f + (1-f)\tilde{\varepsilon}_b. \tag{4}$$

Using this assumption, and h=[0.. 1... 0]′ with 1 at the m-th position [Eqn. 2], we can show that,

$$\Omega = \langle k_m | \tilde{\varepsilon}_0^{-1} | k_m \rangle. \tag{5}$$

Eqn. 4 has to be true for every integer m. The tensor $\tilde{\varepsilon}_0$ in general can be written as $\tilde{\varepsilon}_0 = Q^T \tilde{\varepsilon}_p Q$ [6], where $\tilde{\varepsilon}_p$ is the dielectric tensor of the anisotropic material in its principal coordinate system, Q is a rotational operator (an orthogonal operator), and Q^T is the transpose of Q. With this $\tilde{\varepsilon}_0$, Eqn. 5 can be rewritten using the inner product properties [21] to,

$$\Omega = \langle k_m | Q^T \tilde{\varepsilon}_p^{-1} Q | k_m \rangle = \langle Qk_m | \tilde{\varepsilon}_p^{-1} | Qk_m \rangle, \tag{6}$$

where $Q|k_m\rangle = \mathbf{k}' - \mathbf{G}'_m = [k_{x'} - G_{mx'}, k_{y'} - G_{my'}]'$ is the rotated vector of $|k_m\rangle = \mathbf{k} - \mathbf{G}_m = [k_x - G_{mx}, k_y - G_{my}]'$. If one of the material is assumed to be isotropic, and the other is assumed to be anisotropic which is often the case [6,19], then using Eqn. 4 we have,

$$\tilde{\varepsilon}_p^{-1} = \begin{pmatrix} 1/\{\varepsilon_a f + \varepsilon_{b1}[1-f]\} & 0 \\ 0 & 1/\{\varepsilon_a f + \varepsilon_{b2}[1-f]\} \end{pmatrix}, \tag{7}$$

where $\tilde{\varepsilon}_a$ is taken as an identity matrix multiplied by a constant, $\varepsilon_a = n_a^2$ (i.e., an isotropic material) and $\tilde{\varepsilon}_b$ is assumed as the dielectric tensor of the anisotropic material in the principal coordinate system, with ε_{b1} and ε_{b2} being the principal dielectric constants. Substituting the expressions for $Q|k_m\rangle$ and $\tilde{\varepsilon}_p^{-1}$ in Eqn. 6, we obtain the following equation for the EFS of the H-polarization in the rotated frame,

$$\frac{(k_{x'} - G_{mx'})^2}{\varepsilon_a f + \varepsilon_{b1}[1-f]} + \frac{(k_{y'} - G_{my'})^2}{\varepsilon_a f + \varepsilon_{b2}[1-f]} = \Omega. \tag{8}$$

Eqn. 8 describes an ellipse in the rotated frame ($k_{x'}$, $k_{y'}$) with the origin at $\mathbf{G}'_m = [G_{mx'}\ G_{my'}]'$. As the equation is true for every m, there will be repetitions of the same ellipse for each m [corresponding to each reciprocal lattice vector].

Similarly, for the E-polarization, it can be shown that, the one plane wave approximation leads to $|k_m|^2 = \Omega\varepsilon_0$, where $\beta(m-n) = \delta_{mn}/\varepsilon_0$, has been used for Eqn. 3. Assuming $|k_m\rangle = \mathbf{k} - \mathbf{G}_m$, we will arrive at the same equation as Eqn. 8, however the denominators at the left hand side of the Eqn. 8 are replaced with $\varepsilon_0 = \varepsilon_a f + \varepsilon_{b3}(1-f)$. Here, ε_{b3} is the principal dielectric constant of the anisotropic material experienced by the E-polarization. Note that for the E – polarization Eqn. 8 describes a circle.

To illustrate the EFS construction, let's first assume both materials are isotropic (i.e. $\varepsilon_{b1}=\varepsilon_{b2}=\varepsilon_{b3}=n_b^2$) and consider a 2D PC with a square lattice, n_a=1.6, f=0.4, and a very small dielectric modulation (i.e. $n_a \approx n_b$). Note that, with isotropic materials, and very small dielectric modulation, the distinction between the EFSs of E and H – polarizations vanishes.

For an instance of the EFS construction, consider ω=0.5. Using the radius given by Eqn. 8, let's draw circles at each reciprocal lattice point as shown in Figure 3(a). Each closed contour constitutes to an EFS of a particular band. The bands are indexed according to their positions in the extended zone diagram [14-15].

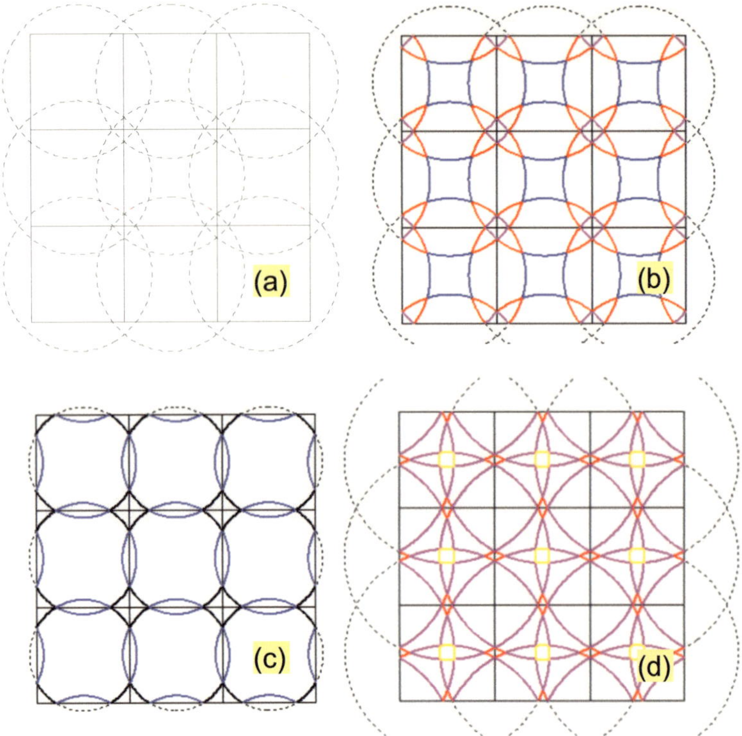

Figure 3. EFS constructions of a square lattice PC with f=0.4, n_a= 1.6 and $n_a \approx n_b$. Band indices: band 1 – black, band 2 – blue, band 3 – red, band 4 – pink, band 5 – yellow. The square boxes indicate boundaries of the BZs (a) EFS construction for ω=0.5 (b) EFS for ω=0.5 (c) EFS for ω=0.36 (d) EFS for ω=0.67.

Now let's demonstrate the validity of the EFS obtained from the one plane wave approximation, by comparing it with the numerically evaluated. In Figures 4(a) and 4(b), we plot EFSs (ω=0.5) for n_b=2.0 and 2.6, respectively, and all other parameters are kept the same as in Fig. 3. The one plane wave construction is shown in the black dashed lines, whereas the EFS obtained from the numerical calculation is highlighted in color. As it is clear from Fig. 4(a), for the PC with weak dielectric modulation (i.e., n_b=2.0), the one plane wave construction agrees well with the numerically calculated EFS. However, when the dielectric modulation increases, the degree of disagreement increases (Fig. 4(b)). Nevertheless, in PCs with large dielectric modulations, one plane wave approximation still can be used to gain a rough idea on the EFSs shapes, for frequencies far from the photonic band edge. It is important to note that one or two plane wave approximations (discussed in Sec. 2.1.2) fail when the frequency is close to the photonic band edge (see Fig. 14; Sec. 4.2; and [2]).

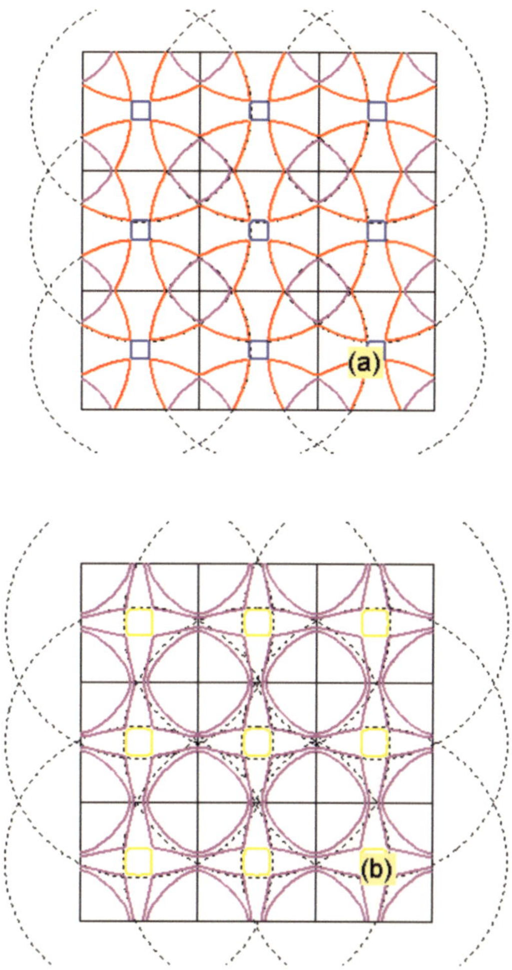

Figure 4. EFS constructions (dashed black line) and the numerically calculated EFS (color) of a square lattice PC with f=0.4 n_a=1.6, and ω=0.5. Band indices: band 3 – red, band 4 – pink, band 5 – yellow. (a) EFS for n_b=2.0 (b) EFS for n_b=2.6

Finally, we provide an EFS construction example for a 2D PC made of an anisotropic material. Consider a square lattice PC with an anisotropic material ($\sqrt{\varepsilon_{b1}}$=1.6, $\sqrt{\varepsilon_{b2}}$=2.0), n_a=1.8, and f=0.2. The principal axis with the dielectric constant ε_{b1} is oriented 45° with respect to the k_x – axis of the PC, as shown Figure 5(a). The Q operator in Eqn. 6 is a 45° anticlockwise rotational operator.

Hence, based on Eqns. 6 – 8, we draw ellipse at each reciprocal lattice point in the rotated coordinate system as shown in Fig. 5(a). The numerically calculated EFS for the H–polarization with ω=0.5 is shown in Fig. 5(b), and as we can readily verify from the figure, both the construction and numerical evaluation share a good agreement.

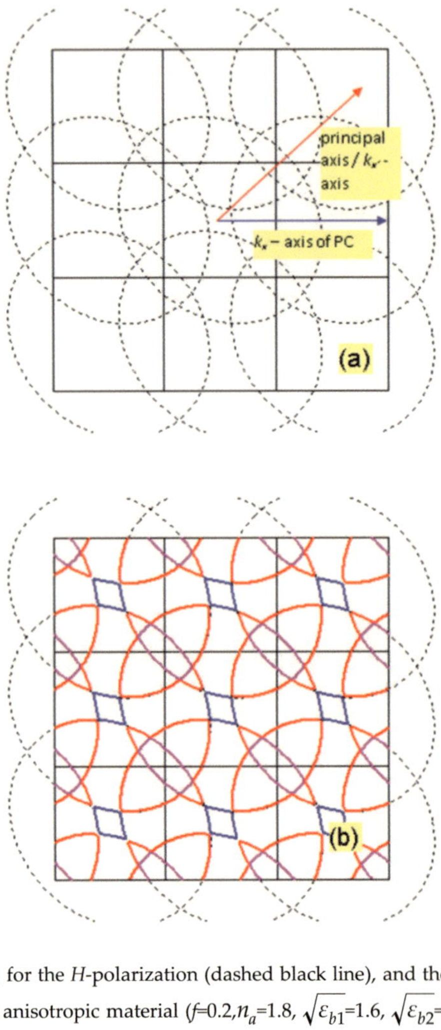

Figure 5. EFS construction [ω=0.5] for the H-polarization (dashed black line), and the numerically calculated EFS (color) of a square lattice PC with an anisotropic material (f=0.2,n_a=1.8, $\sqrt{\varepsilon_{b1}}$=1.6, $\sqrt{\varepsilon_{b2}}$=2.0). Band indices: band 2 – blue, band 3 – red, band 4 – pink. (a) EFS construction (b) EFS with the band index assignment.

2.1.2. *Two plane wave approximation*

Although, one plane wave approximation seems to be good in approximating the EFS of a 2D PC with a weak dielectric modulation, a magnified version of the one plane wave EFS, will show that this approach is unable to approximate the edges of the EFS accurately. This problem can be addressed by using two plane waves approximation.

For the sake of simplicity, we will demonstrate the two plane waves approximation for 2D PCs made of isotropic materials. We will further assume the dielectric modulation is finite and weak, such that only two Fourier coefficients of $\beta(i)$, $\beta(0)$ and $\beta(1)$, are significant. With these, we can approximate Eqn. 2 as,

$$\Omega = h_i M_{i,i} h_i + h_i M_{i,i+1} h_{i+1} + h_{i+1} M_{i+1,i} h_i + h_{i+1} M_{i+1,i+1} h_{i+1}, \tag{9}$$

which has to be true for every i. We can choose i=1, and the eigenvalues of the resulting M, can be obtained by finding the determinant of the matrix,

$$\begin{bmatrix} M_{11} - \Omega & M_{12} \\ M_{21} & M_{22} - \Omega \end{bmatrix}. \tag{10}$$

Using the fact $M_{ij}=M_{ji}$, and denoting $\beta(i)$ as β_i, we can show,

$$2\Omega = \beta_0 (k_0^2 + k_1^2) \pm \sqrt{\beta_0^2 (k_0^2 - k_1^2)^2 + (2\beta_1 k_0 k_1)^2}, \tag{11}$$

where $k_i k_j$ represents $<k_i | k_j>$ and $|k_i||k_j|$ for the H-and E-polarizations, respectively. Note that for a weak dielectric modulation, $\beta_0 >> |\beta_1|$ and $\beta_0 > 0$.

Eqn. 11 can be evaluated using specific choices of wavevectors and reciprocal lattice vectors. For an instance, let's examine EFSs near the first band gap of the E-polarization, in the square lattice PC around the X($-\pi/a$,0) point. With $g=2\pi/a$, $\mathbf{k}=[k_x\ k_y]'$, $\mathbf{G}_0=[0\ 0]'$, and $\mathbf{G}_1=[-g\ 0]'$, the terms in Eqn. 11 become,

$$\begin{aligned} \beta_0 (k_0^2 + k_1^2) &= 2\beta_0 [(k_x + g/2)^2 + k_y^2 + g^2/4] \\ \beta_0^2 (k_0^2 - k_1^2)^2 &= [2\beta_0 g (k_x + g/2)]^2 \\ (2\beta_1 k_0 k_1)^2 &= 4\beta_1^2 [(k_x + g/2)^2 + k_y^2 + g^2/4]^2 - [2\beta_1 (k_x + g/2)]^2. \end{aligned} \tag{12}$$

Let's move the origin of the reciprocal space from (0,0) to the X point, by writing $k_X=k_x+g/2$. In the new coordinate system, using Eqns. 11 and 12, we can show that,

$$\Omega = \beta_0 (k_X^2 + k_y^2 + g^2/4) \pm \sqrt{(\beta_0 k_X g)^2 + \beta_1^2 (k_X^2 + k_y^2 + g^2/4)^2 - (\beta_1 k_X g)^2}. \tag{13}$$

Eqn. 13 does not provide any insight to the shape of the EFS. More useful information can be obtained if we can transform the equation to a simpler form. To do this, we first seek an approximation to the square root term in Eqn. 13, for a small k_X. For a small k_X, the square root term in Eqn. 13, can be written as $y = \sqrt{P^2 + Q^2}$, where $P = |\beta_0 k_X g|$ and $Q = \beta_1 (k_X^2 + k_y^2 + g^2/4)$ (the term $\beta_1 k_X g$ in Eqn. 13 is neglected). We can approximate y using a binomial expansion as $\approx Q + \frac{P^2}{2Q}$. However, this requires $Q >> P$, and in fact this is a good assumption as we shall justify later in the discussion. Using the binomial expansion of y, for a small k_X and k_y, Eqn. 13 can be simplified to,

$$\left[\Omega_1 - \frac{g^2}{4}(\beta_0 - |\beta_1|)\right] = k_X^2\left(\beta_0 - |\beta_1| - \frac{2\beta_0^2}{|\beta_1|}\right) + k_y^2(\beta_0 - |\beta_1|), \tag{14}$$

$$\left[\Omega_2 - \frac{g^2}{4}(\beta_0 + |\beta_1|)\right] = k_X^2\left(\beta_0 + |\beta_1| + \frac{2\beta_0^2}{|\beta_1|}\right) + k_y^2(\beta_0 + |\beta_1|), \tag{15}$$

where the two solutions of Eqn. 13 are denoted as Ω_1 and Ω_2. These two solutions correspond to frequencies in band 1 [Ω_1] and band 2 [Ω_2]. With $\beta_0 > 0$ and $\beta_0 > |\beta_1|$, we have $\beta_0 + |\beta_1| + \frac{2\beta_0^2}{|\beta_1|} > 0$, in Eqn. 15. Therefore, EFSs for band 2 are elliptical (provided that $Q >> P$) with the lengths of the semi –major and –minor proportional to the reciprocals of $\sqrt{\beta_0 + |\beta_1| + \frac{2\beta_0^2}{|\beta_1|}}$ and $\sqrt{\beta_0 + |\beta_1|}$, respectively. On the other hand, note that, the EFSs for band 1 (Eqn. 14, with $\beta_0 - |\beta_1| - \frac{2\beta_0^2}{|\beta_1|} < 0$) are not elliptical.

Now let's examine the validity of the assumption $Q > P$ in deriving Eqns. 14 and 15. The expressions P and Q are linear and quadratic functions of k_X, respectively, and the minimum point of Q is $f(k_y) = |\beta_1|\left(k_y^2 + \frac{g^2}{4}\right)$. Hence, a sufficient condition for $Q > P$ is simply $f(k_y) > P$. Inserting the expression for P, this condition becomes,

$$-\frac{|\beta_1|}{\beta_0 g}(k_y^2 + \frac{g^2}{4}) < k_X < \frac{|\beta_1|}{\beta_0 g}(k_y^2 + \frac{g^2}{4})$$

The conditions on k_y can be obtained by solving the quadratic in-equality $Q - P > 0$. Solving this in-equality, we have,

$$k_y > \frac{g}{2|\beta_1|}\sqrt{\beta_0^2 - \beta_1^2}.$$

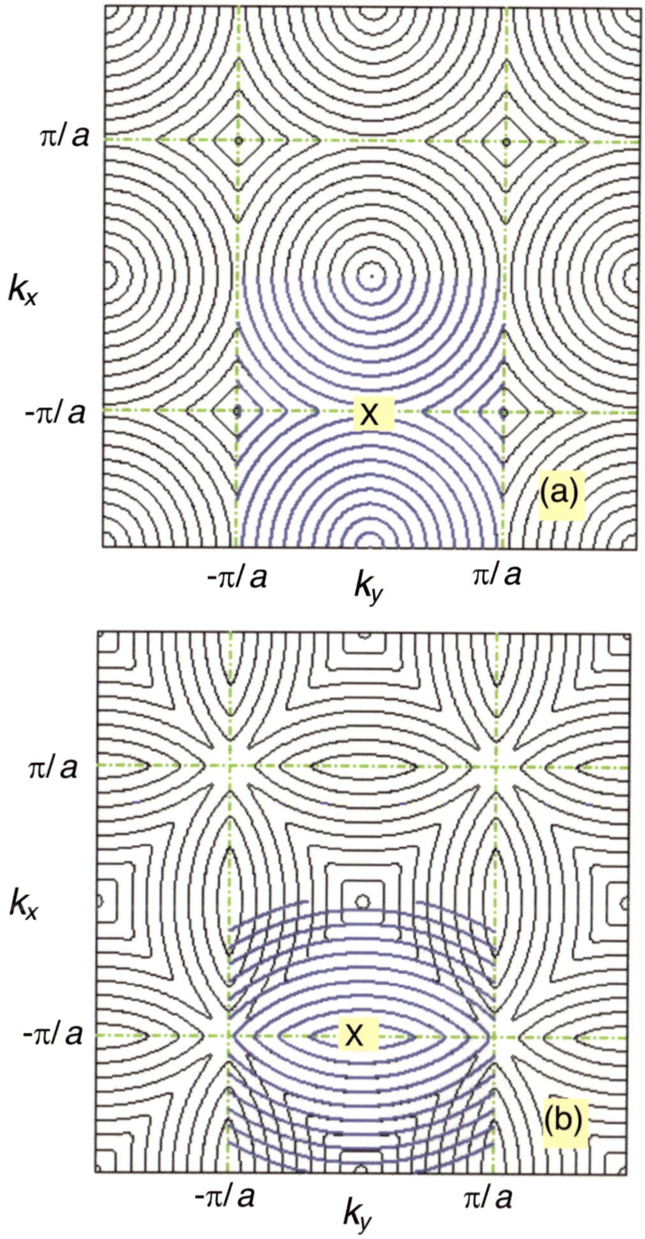

Figure 6. EFS approximations using two plane waves approximation [square lattice PC with n_a=1.6, n_b=1.8, and f=0.2]. The numerically calculated EFS (thin black lines) and the approximated EFS (thick blue lines) for frequencies in the (a) first band (b) second band.

To illustrate the EFS construction using the two plane waves approximation, consider a square lattice PC with n_a=1.6, n_b=1.8, and f=0.2. The numerically calculated EFS and the approximated EFS, for frequencies in the first and the second band are shown in Figures 6(a) and 6(b), respectively. In both figures, we have evaluated Eqn. 11 for-$\pi/a < k_y < -\pi/a$ and-$2\pi/a < k_x < 0$, and the results share a good agreement with those of numerical evaluation. As Eqn. 9 has to be true for every i, we can translate the EFS approximated for i=1 by a vector $\mathbf{G}_i$. Furthermore the EFSs in the neighborhood at the other symmetrical X points [i.e. (π/a,0), (0,π/a) and (0,-π/a)] can be obtained using the EFS approximated in the neighborhood of X(-π/a,0) by point group – symmetry operations [9, 22].

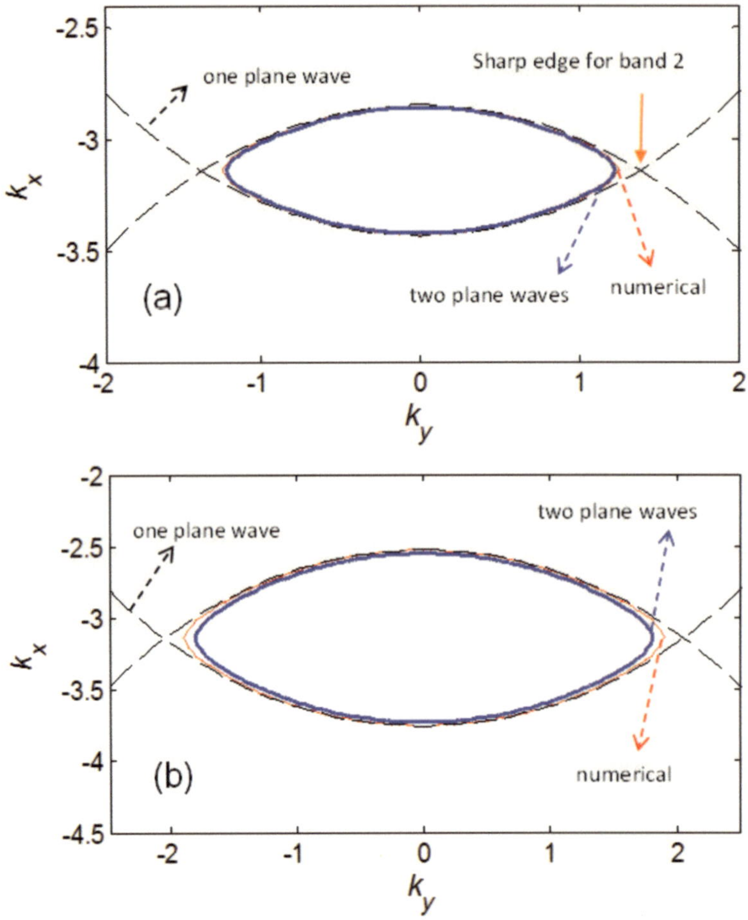

Figure 7. The merit of the two plane waves approximation. EFSs for a square lattice PC with n_a=1.6, f=0.2. EFS (ω=0.31) obtained from the one plane wave approximation (dashed black line), two plane waves approximation (thick blue line), and the numerical calculation (thin red line), for (a) n_b=1.8 (b) n_b=2.0. Both k_x and k_y in (a) and (b) have units of $1/a$.

In order to appreciate the ability and inability of the two plane waves approximation, we plot EFSs (band 2, ω=0.3) obtained from the one plane wave approximation, two plane waves approximation, and the full numerical calculation in Figures 7(a) and 7(b), for n_b=1.8 and n_b=2.0, respectively. The one plane wave – EFS is constructed using two circles originated from (0,0) and (-g, 0) points. Thus, it always predicts sharp edges for the EFS (see band 2 in Figs. 3(c), 7(a) and 7(b)). The sharp edges does not appear in the numerically calculated EFS, which is the accurate EFS. The failure of the one plane wave method is corrected in the two plane waves approximation. The two plane waves approximation perfectly matches the numerical counterpart when the dielectric modulation is small (Fig. 7(a)). As expected, when the dielectric modulation becomes large, the two plane waves approximation becomes poorer (Fig. 7(b)), however the approach still exhibits a better accuracy compared to the one plane wave approach.

3. Propagation directions

When light crosses the boundary between two homogenous mediums (medium 1 and medium 2), light refracts. The light in medium 1 with an incident angle, θ_i, can excite two symmetrical waves (with angles θ_p; see Figure 8) in medium 2. What happens to this refraction picture, when we replace the homogenous dielectric medium 2, with a PC?. When we replace the homogenous medium 2, with a PC, the simple refraction picture based on the Snell's law will disappear. A more general technique has to be used, in order to find light propagation directions. This section reviews the well – known method [1-4] of finding light propagation directions in PCs, based on their EFS. As a pre-requisite for the second part in this section, we will first give a brief review on anisotropic PCs [22].

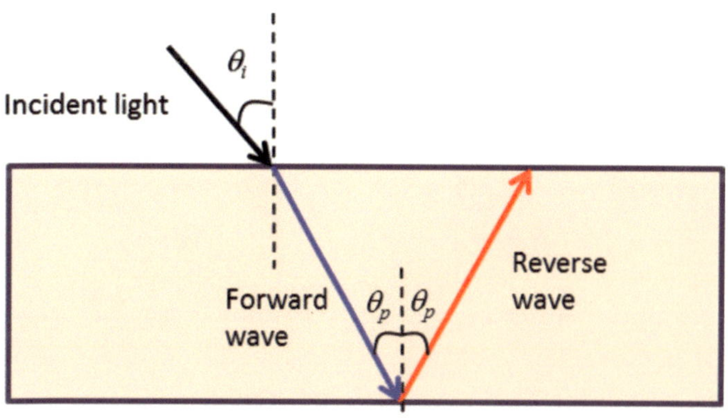

Figure 8. Light propagation from a homogenous dielectric medium 1 to a homogenous dielectric medium 2

3.1. Orientation parameter — Anisotropic PC

A PC may be simply viewed as a lattice with a motif attached to each lattice point. If the re-orientation of the motif causes the symmetry elements of the PC to change, then the corresponding PC is an anisotropic PC, as opposed to an isotropic PC, where the orientation of the motif is irrelevant to the symmetry of the PC. A 2D PC made of only isotropic dielectric materials, with a circular motif in a 2D lattice, is a clear example of a 2D isotropic PC. Anisotropic PCs, on the other hand, can be geometrically anisotropic or materially anisotropic. Figure 9(a) shows the geometry of a 2D hexagonal lattice PC with a square motif. The orientation of the square motif with respect to the underlying lattice plays a crucial role in determining the optical properties of the 2D PC. In Fig. 9(b), instead of a square motif, we have a circular motif for which the corresponding orientation is irrelevant. If all materials are isotropic, then the geometries in Figs. 9(a) and 9(b) represent examples of geometrically anisotropic and isotropic PCs, respectively. In the presence of an anisotropic material, the optical properties of the PC will vary in accordance to the orientation of the anisotropic material (i.e., the orientation of the principal axes with respect to the lattice), and therefore the corresponding PC is defined as a materially anisotropic PC. A PC with a geometry in Fig. 9(b) constitutes to a materially anisotropic PC, if either the matrix or the circular cylinder is an anisotropic medium. On the other hand, in the presence of an anisotropic material, a PC with a geometry as shown in Fig. 9(a) has a mixed anisotropy (i.e., both geometrical and material's anisotropy exist).

As we will show in the next section, the orientation parameter of the anisotropic PC is crucial, as to determine the accurate light propagation direction.

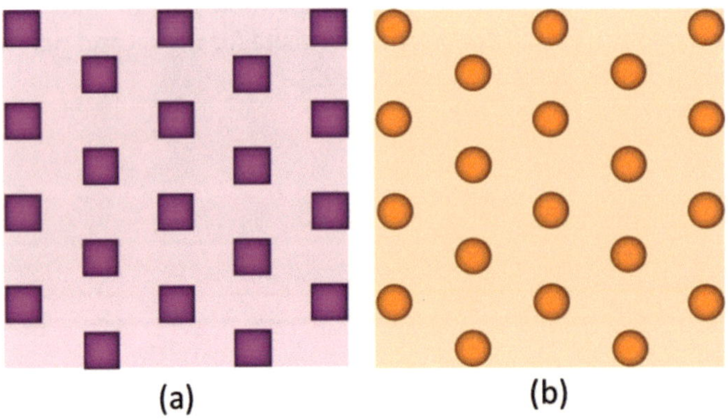

Figure 9. Examples of geometries of the 2D hexagonal lattice PC. (a) square motif (b) circular motif

3.2. Technique of determining light propagation direction

The direction of light propagation in any medium is given by the direction of the group velocity, v_g. Group velocity is defined as $v_g = \nabla_{\mathbf{k}}\omega$, where $\nabla_{\mathbf{k}}$ is the gradient operator in the wavevector space. The light propagation direction at the frequency, ω, can be determined using the gradient of the corresponding EFS.

Consider the problem of light of propagation from medium 1 (m_1) to medium 2 (m_2), as in Fig. 8. Assume m_1 to be a bulk isotropic dielectric material. For a general case, assume m_2 to be a 2D anisotropic PC, and for the sake of discussions, assume the anisotropic PC to be materially anisotropic. In finding the light propagation directions in m_2, there are two important orientations. These are:

- Motif orientation – All anisotropic PC have an orientation parameter (see Sec. 3.1). Let's define the orientation parameter to be an angle, α. This angle defines the orientation of the motif with respect to the normal line of the interface.
- Lattice orientation – EFS of m_2 is usually calculated based on a coordinate system that is aligned with one of the symmetrical axis of the lattice system [based on which the PC is constructed; examples in 2D include square lattice, rectangular lattice, honeycomb lattice, and hexagonal lattice]. This coordinate system does not necessarily align with the interface (i.e., the boundary between m_1 and m_2). This misalignment is taken into account by defining a lattice angle, ϕ.

In Figure 10, we illustrate the definitions of α and ϕ via an example.

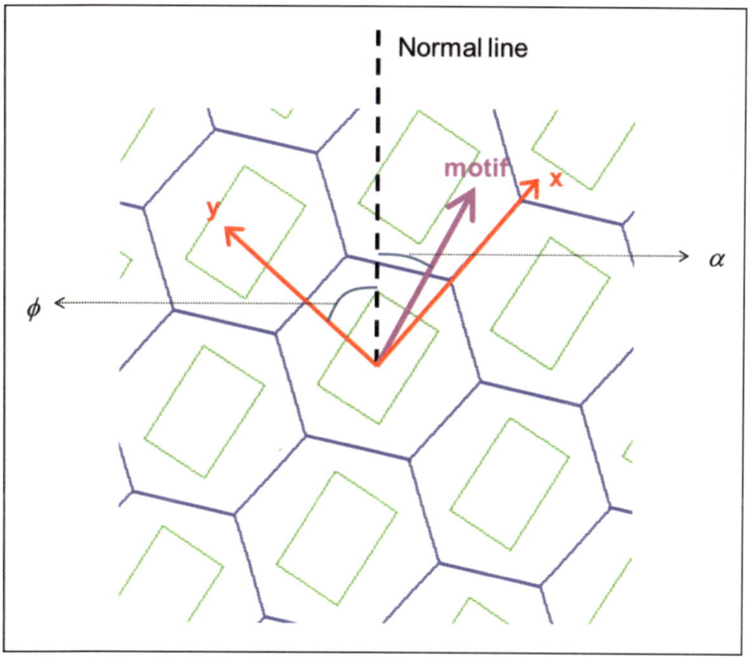

Figure 10. An example of m_2: A 2D hexagonal lattice PC with rectangular motifs [7]. The figures displays a microscopic view, with the details of the motif arrangement and orientation. The motifs are shown in the dark green color, while the unit cell constructions are in the blue color. The interface is normal to the dashed line (i.e., the normal line). The angle of the symmetrical axis of the motif (pink arrow) with respect to the normal line is defined as the angle of motif orientation. The angle between the symmetrical axis of the lattice (red arrows) and the normal line is defined as the lattice angle.

Figure 11(a) shows momentum space diagram that contains the EFS of the two mediums. The angles α and ϕ are shown in this diagram. EFS of m_1 is a circle, while EFS of m_2 is arbitrarily assumed. When light crosses the boundary between the two mediums, the transversal momentum (i.e., momentum component that is parallel to the interface) of light is conserved. Hence a line of momentum conservation, being a vertical line parallel to the normal line (green dashed), determined by the incident angle, θ_i, and passing through the EFS of m_2 produces two intersection points. The directions of the normal vector at the intersection points between the EFS of m_2, and the line of momentum conservation give the propagation directions in m_2. The normal vector with the sign of the vertical component same as the sign of the incident wavevector'svertical component is considered, as a forward wave and the other, is considered as a reverse wave [Fig. 11(a)] [23]. In Fig. 11(a), the sign of $\nabla_k \omega$ is assumed as negative. However, take note that, depending on the sign of $\nabla_{\mathrm{k}} \omega$, the gradient can be either inward or outward. The consequence of such changes in the sign of $\nabla_{\mathrm{k}} \omega$ to the directions of forward and reverse waves is illustrated in Figure 11(b).

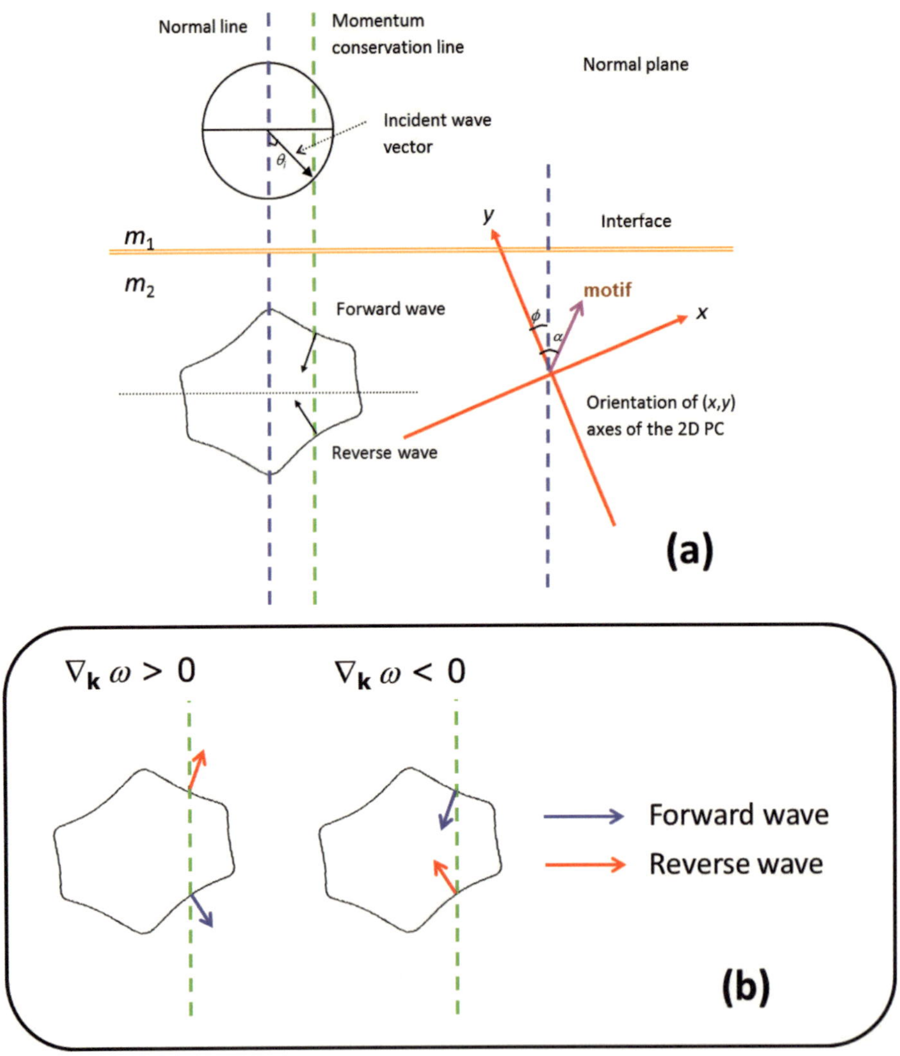

Figure 11. (a) In-plane propagation of light from m_1 to m_2. m_1: isotropic bulk medium, m_2: anisotropic PC (b) The sign of the group velocity, and the directions of the forward and reverse waves

http://dx.doi.org/10.5772/59887

4. Applications of the EFS analysis

In this section we will use EFSs to describe few of the peculiar phenomena observed in PCs. Specifically, we will apply the concept of EFS to describe superprism, effective negative index mediums, negative refractions, and superlenses in PCs.

4.1. Superprism effect

Superprism effect was first demonstrated in the PC by Kosaka *et al.* [1,24]. As the name suggests, a superprism is a special and a superior version of the ordinary prism. The superprism has an extraordinary sensitivity to the incident wavelengths [24], and incident angles [1].

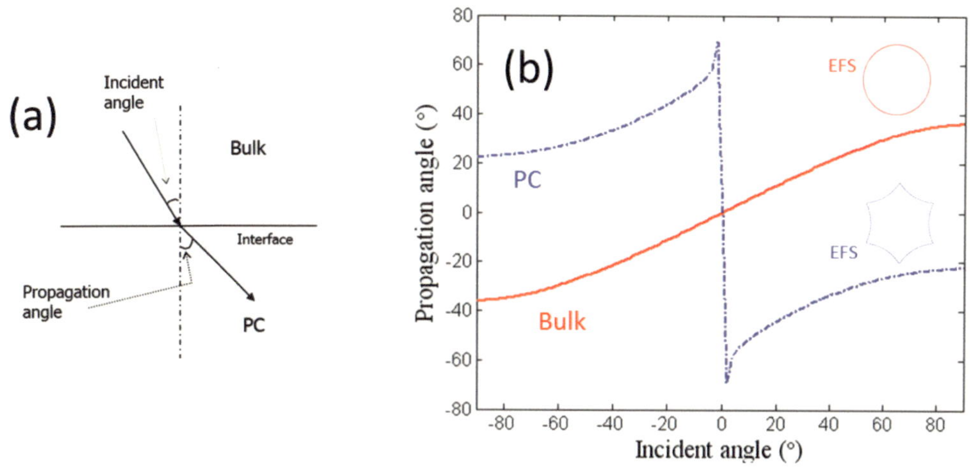

Figure 12. Angle – sensitive superprism effect

The angle sensitive superprism effect can be easily understood. In Figure 12, we show an example of the propagation angle versus incident angle plot at a specific frequency. The parameters for this plot can be found in [12]. The red curve corresponds to a bulk medium, while the blue curve is for a 2D hexagonal lattice PC. For the bulk medium, the propagation angle follows the Snell's Law, and the relationship between the incident angle and propagation angle is smooth (red curve). On the other hand, for the 2D PC, near the normal incidence, the propagation angles become highly sensitive to the incident angles (blue curve). Such a high sensitivity is due the sharp edge observed in the corresponding hexagonal like EFS of the 2D PC [see the blue EFS in the insert].

The wavelength dependent superprism effect is useful for the spatial filtering of multiple wavelengths. Therefore, it is very useful for implementation of a compact arrayed waveguide grating [24]. The operation principle of the wavelength dependent superprism effect is graphically shown in Figure 13. Fig. 13(a) shows an ordinary prism, while Fig. 13(b) shows a superprism.

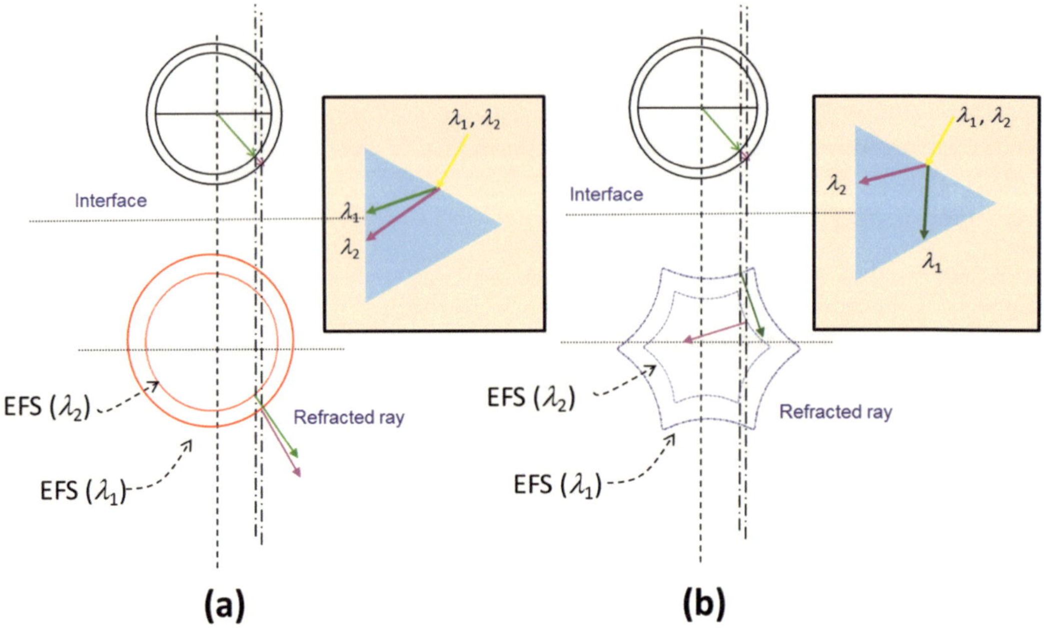

Figure 13. Wavelength dependent superprism effect. (a) Ordinary prism; (b) Superprism [Note that the signs of $\nabla_{\mathbf{k}}\,\omega$ for the red and blue EFSs, are positives and negatives, respectively.]

Let's consider two closely spaced wavelengths, λ_1 and λ_2. The EFSs for these two wavelengths are circles of different sizes in a bulk medium and in an ordinary prism. However, in a PC the EFSs can be very different. For superprism effects EFSs with sharp edges are preferred. By plotting the EFSs of the PC at various normalized frequencies (a/λ), EFSs with sharp edges can be identified. Then by adjusting a, we can design the superprism at the desired wavelengths. In Fig. 13(b), example of EFSs in a 2D hexagonal lattice PC are shown. These EFSs have sharp edges, and the EFSs have different sizes. By selecting the incident angle appropriately, we can design the corresponding EFSs gradient vectors of the two wavelengths, to be at two different curvatures. As shown in Fig. 13(b), this will cause a huge difference in the propagation angles of the two closely spaced wavelengths.

4.2. Effective negative index mediums

The behavior of an electron near the band edge of a semiconductor can be approximated as that of a free electron with an effective mass. The analogous concept in a PC was first shown by Notomi [2].

The EFS of a 2D PC with a symmetrical unit cell (i.e. the unit cell have a third or higher order rotational axes), becomes circular at the bottom edge of the first band (i.e., in the long wavelength limit) and for such a frequency range, the PC behaves like a bulk isotropic medium with an effective refractive index [25-26]. The effective refractive index (n_{eff}) can be assigned by fitting the EFS to the expression $\omega = k/|n_{eff}|$ [note that, ω and k are normalized]. The effective index for the first band is positive, since the first band has a positive sign of $\nabla_{\mathbf{k}}\omega$. It is also reported that in such 2D PCs, when the modulation of the refractive index become stronger, the EFSs become circular for frequencies near the edges of the higher order bands. Thus for the corresponding range of frequencies, the 2D PC can be homogenized to an isotropic bulk medium with an effective refractive index. However, this effective index can be negative if the sign of $\nabla_{\mathbf{k}}\omega$ of the particular band is negative [2].

For an instance, a 2D photonic band structure of the hexagonal lattice PC with circular rods, and huge refractive index contrast is shown in Figure 14 (only the first two bands are shown) [2]. As we can see from this figure, the EFSs close to the bottom edge of the first band, and the top edge of the second band are circular. As frequency increases, the EFSs of the first band move outwards (increasing in size). Thus, the first band has a positive $\nabla_{\mathbf{k}}\omega$, and the corresponding circular EFSs are defined with effective positive refractive indices. On the other hand, in the second band, as frequency increases the EFSs move inwards (decreasing in size). Hence, the second band has a negative $\nabla_{\mathbf{k}}\omega$, and therefore their circular EFSs have effective negative refractive indices.

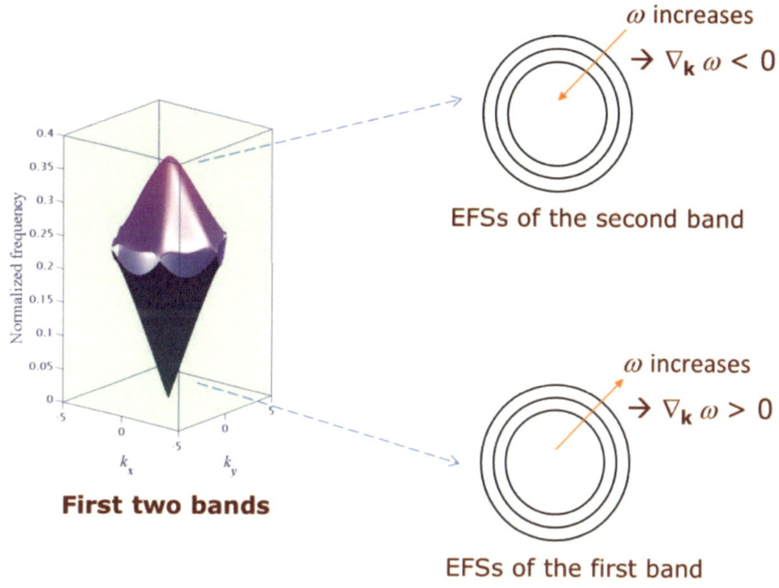

Figure 14. Example of circular EFSs of a PC with positive and negative effective refractive indices.

How to obtain effective anisotropic materials using PC?. Such investigations may lead to novel polarization splitting, and tunable devices.

It has been shown that, 2D materially anisotropic PCs, with symmetrically no rotational axes of order larger than two, exhibit elliptical EFSs in the long wavelength limit [25-26]. Near the higher order band edges, they do behave like a bulk anisotropic media, and if $\nabla_{\mathbf{k}}\omega<0$, a set of effective negative principle refractive indices can be defined. The effective principal refractive indices can be defined by fitting to the expression, $\frac{k_x^2}{n_{p1}^2} + \frac{k_y^2}{n_{p2}^2} = \omega^2$, where n_{p1} and n_{p2} are the effective principal refractive indices (note that ω, k_x, and k_y are normalized). As an example, in Figure 15, we show the 2D photonic band structure and the EFSs (second band) for a 2D hexagonal lattice PC with rectangular motif [7]. From this figure, we can see that, near the top edge of the second band, the EFSs are elliptical, and the signs of $\nabla_{\mathbf{k}}\omega < 0$ (i.e., negative principal refractive indices).

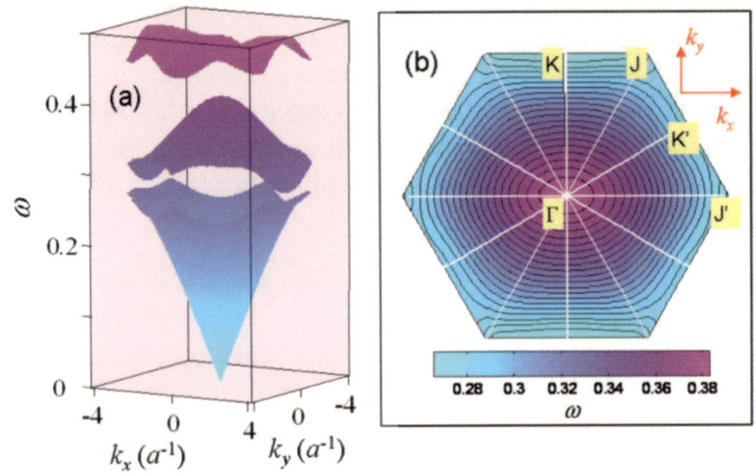

Figure 15. (a) 2D photonic band structure of the anisotropic PC – rectangular motif in 2D hexagonal lattice. For the details of the structure, please see [7]. (b) The contour plot of the second band in (a);

4.3. Negative refractions and superlenses

In the usual (positive) refraction, when light crosses the interface of the two mediums (Fig. 8), light bends on the opposite sides of the normal line. However, if light bends on the same side of the normal line, then the corresponding bending is defined as a negative refraction. Negative refraction was first predicted by Veselago in 1968 [27]. Negative refractions have been observed in artificial mediums including PCs [28-30].

For an instance of negative refraction in the PC, take a look on Figs. 12(a) and 12(b). Both incident and propagation angles in the configuration of Fig. 12(a) are defined to be positive values. If the incident (or the propagation) angle goes the other side of the normal, then it is negative. Using this formalism, it easy to recognize that the red curve (corresponds to the bulk medium) shows propagation angles for the positive refractions, and the blue curve (corresponds to the PC) shows propagation angles for the negative refractions.

An important application of the negative refraction is a construction of a superlens [31]. Negative refraction alone does not satisfy the requirement to build a superlens. The corresponding structure also should exhibit an effective negative refractive index. Note that, a

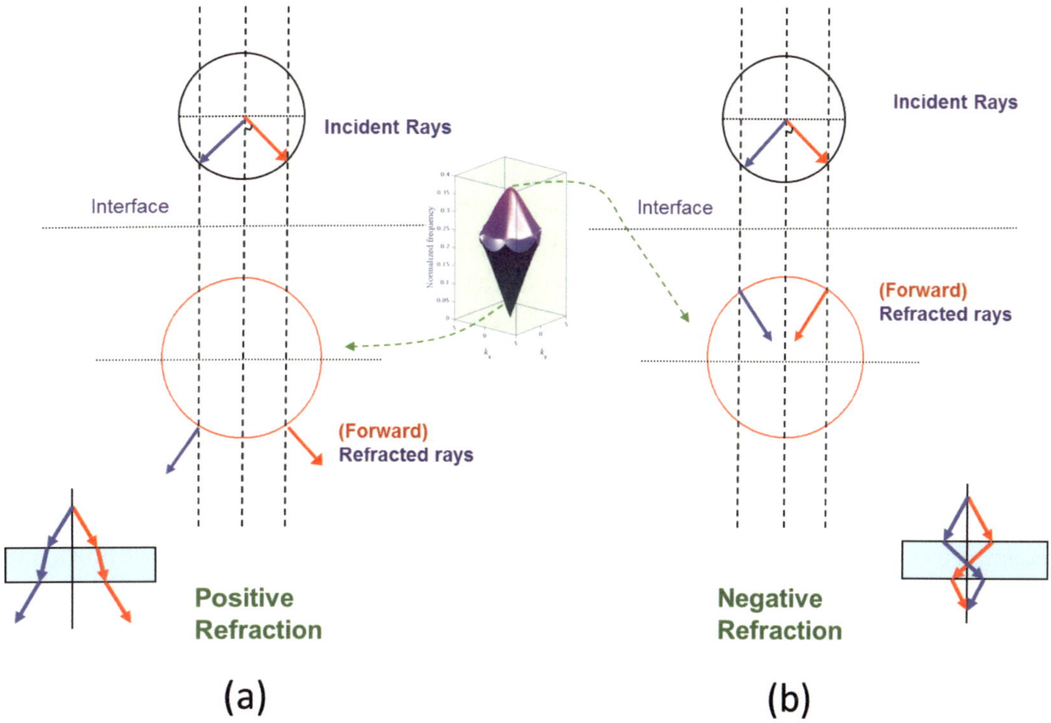

Figure 16. Rays from a point object (a) passing through a positive refractive index medium (b) passing through a negative refractive index medium. In both (a) and (b), the reflected rays are not shown.

negative refraction does not guarantee a negative refractive index. A effective negative refractive index (n_{eff}) can be only defined when the EFS is circular (see Sec. 4.2). In Fig. 12(b), although we see negative refractions, the corresponding EFS (blue color) is not circular. Thus, in this case a negative refractive index cannot be defined. However, negative refractive indices can be defined for frequencies close to the top edge of the second band in Fig. 14. This frequency region has circular EFSs with $\nabla_k \omega < 0$.

Superlens has flat surfaces, and they produce images with subwavelength resolutions. The details on the subwavelength resolution of the superlens can be found in [31]. Here, we would like to explain, how the flat surface can result in a focusing effect. The focusing effect of the superlens is graphically illustrated in Figure 16.

Consider a homogenous slab (as in Fig. 8), and a point object at some distance from the slab in an air ambience. Firstly, assume the slab has a positive refractive index (as for the bottom edge frequencies of the first band in Fig. 14). As shown in Fig. 16, the usual positive refraction in the slab, will cause the rays from the point object to diverge. If the slab has negative refractive index (as for the top edge frequencies of the second band in Fig. 14), the negative refractions will bring the rays from the point object to a focus (see the illustration in Fig. 16(b)).

5. Summary

EFS is the surface resulting from the projection of the photonic band structure onto the wavevector space, at a constant frequency. In PCs, EFSs can be accurately modeled using a plane wave expansion method. For 2D PCs with weak dielectric modulation, EFSs can be obtained using one and two plane wave approximation techniques [Sec. 2]. Though one plane wave approximation succeed to a large extend in predicting the EFS of a 2D PC with a weak dielectric modulation, the approximation is inaccurate for wavevectors near the boundaries of the BZ. The one plane wave approach often predicts a very sharp edge to the EFS. The deficiencies in the one plane wave approximation is corrected in the two plane waves approach.

EFS is an essential tool in determining light propagation directions in the PC. The gradient of the EFS gives information about the group velocity direction and the sign of the propagation angle (Sec. 3). EFSs can be used to analyze various peculiar light propagations in the PC. Examples that cover superprism effects, negative index mediums, negative refractions, and superlens are discussed in Sec. 4.

Author details

G. Alagappan*

Address all correspondence to: gandhi@ihpc.a-star.edu.sg

Department of Electronics and Photonics, Institute of High Performance Computing (IHPC), Agency for Science, Technology and Research (A-STAR), Singapore

References

[1] Kosaka H, Kawashima T, Tomita A, Notomi M, Tamamura T, Sato T, Kawakami S. Superprism phenomena in photonic crystals. Physical Review B 1998; 58(16) R10096-R10099.

[2] Notomi M. Theory of light propagation in strongly modulated photonic crystals: Refractionlike behavior in the vicinity of the photonic band gap. Physical Review B 2000; 62 (16) 10696-10705.

[3] Foteinopoulou S, Soukoulis CM. Electromagnetic wave propagation in two-dimensional photonic crystals: A study of anomalous refractive effects. Physical Review B 2005; 72(16) 165112.

[4] Jiang W, Chen RT, Lu X. Theory of light refraction at the surface of a photonic crystal. Physical Review B 2005; 71(24) 245115.

[5] Alagappan G, Sun XW, Yu MB. Equal-frequency surface analysis of two-dimensional photonic crystals. Journal of Optical Society of America A 2008; 25(1) 219-224.

[6] Alagappan G, Sun XW, Shum P, Yu MB, Engelsen D. Symmetry properties of two-dimensional anisotropic photonic crystal. Journal of Optical Society of America A 2006; 23(8) 2002-2013.

[7] Alagappan G, Sun XW, Yu MB. Negative principal refractive indices and accidental isotropy in two dimensional photonic crystals with an asymmetrical unit cell. Physical Review B 2007; 76(16) 165123.

[8] Ho KM, Chan CT, Soukoulis CM. Existence of a photonic gap in periodic dielectric structures. Physical Review Letters 1990; 65(25) 3152-3155.

[9] Sakoda K. Optical Properties of Photonic Crystals. Berlin: Spinger; 2005.

[10] Li MS, Wu ST, Fuh AYG. Superprism phenomenon based on holographic polymer dispersed liquid crystal films. Applied Physics Letters 2006; 88(9) 91109.

[11] Baumberg JJ, Perney NMB, Netti MC, Charlton MDC, Zoorob M, Parker GJ. Visible-wavelength super-refraction in photonic crystal superprisms. Applied Physics Letters 2004; 85(3) 354-356.

[12] Alagappan G, Sun XW, Shum P, Yu MB. Tunable superprism and polarization splitting in a liquid crystal infiltrated two-dimensional photonic crystal made of silicon oxynitride. Optics Letters 2006; 31(8) 1109-1111.

[13] Liu YJ, Sun XW. Electrically tunable two-dimensional holographic photonic crystal fabricated by a single diffractive element. Applied Physics Letters 2001; 89(17) 171101.

[14] Harrison WA. Fermi Surface in Aluminum. Physical Review 1959; 116(3) 555-561.

[15] Ziman JM. Principles of the Theory of Solids. Cambridge: Cambridge University Press; 1972.

[16] Busch K, John S. Photonic band gap formation in certain self-organizing systems. Physical Review E 1998; 58(3) 3896-3908.

[17] Busch K, John S. Liquid-Crystal Photonic-Band-Gap Materials: The Tunable Electromagnetic Vacuum. Physical Review Letters 1999; 83(5) 967-970.

[18] Liu CY, Chen LW. Tunable band gap in a photonic crystal modulated by a nematic liquid crystal. Physical Review B 2005; 72(4) 045133.

[19] Alagappan G, Sun XW, Shum P, Yu MB. Engineering the band gap of a two-dimensional anisotropic photonic crystal. Journal of Optical Society of America B 2006; 23(7) 1478-1483.

[20] Takeda T, K. Yoshino. Tunable refraction effects in two-dimensional photonic crystals utilizing liquid crystal. Physical Review E 2003; 67(5) 056607.

[21] Gasiorowicz S. Quantum Physics. John Willey & Sons; 2003.

[22] Alagappan G, Sun XW, Sun HD. Symmetries of the eigenstates in an anisotropic photonic crystal. Physical Review B 2008; 77(19) 195117.

[23] Baba T. Photonic crystal light deflection devices using the superprism effect. IEEE Journal of Quantum Electronics 2002; 38(7) 909-914.

[24] Kosaka H, Kawashima T, Tomita A, Notomi M, Tamamura T, Sato T, Kawakami S. Photonic crystals for micro lightwave circuits using wavelength-dependent angular beam steering. Applied Physics Letters 1999; 74(10) 1370-1372.

[25] Halevi H, Krokhin AA, Arriaga A. Photonic Crystal Optics and Homogenization of 2D Periodic Composites. Physical Review Letter 1999; 82(4) 719-722.

[26] Krokhin AA, Halevi P, Arriaga J. Long-wavelength limit (homogenization) for two-dimensional photonic crystals. Physical Review B 2002; 65(11) 115208.

[27] Veselago VG. The electrodynamics of substances with simultaneously negative values of ε and μ. Soviet Physics Uspekhi (1968);10(4) 509–514.

[28] Shelby RA, Smith DR, Schultz S. Experimental Verification of a Negative Index of Refraction. Science 2001; 292(5514) 77-79.

[29] Berrier A, et al. Negative Refraction at Infrared Wavelengths in a Two-Dimensional Photonic Crystal. Physical Review Letters 2004, 93(7) 073902 (2004).

[30] Schonbrun E, et al. Negative-index imaging by an index-matched photonic crystal slab. Physical Review B 73(19), 195117 (2006).

[31] Pendry JB. Negative Refraction Makes a Perfect Lens. Physical Review Letters 2000; 85(18) 3966-3969.

Chapter 6

Coherent Control of Stimulated Emission Inside one Dimensional Photonic Crystals — Strong Coupling Regime

Alessandro Settimi

Additional information is available at the end of the chapter

http://dx.doi.org/10.5772/59900

1. Introduction

It is well known that emission processes reflect certain inherent properties of atoms, but it has also been demonstrated, in both theory [1,2] and experimentation [3-8], that these same processes are also sensitive to incidental boundary conditions. One example is how they can be modified if contained inside a cavity of dimensions comparable to the emitted light wavelength. The modification can involve emission enhancement or inhibition and is a result of an alteration of field mode structure inside the cavity compared to free space, which can be explained in terms of an interaction between atom and cavity modes [9,10].

The density of states (DOS) can be interpreted as a probability density of exciting a single eigenstate of the electromagnetic (e.m.) field. When the plot of DOS vs. frequency over atomic transition spectral range is found to be smooth, then the rate of emission can be defined by Fermi's golden rule. However, emission dynamics can be drastically modified by photon localization effects [11] and sudden changes in DOS. Such modifications can be interpreted as long-term memory effects and examples of non-Markovian atom-reservoir interactions.

Marked transformations can be induced in the DOS using photonic crystals. These dielectric materials exhibit very noticeable periodic modulations in their refractive indices which result in the formation of inhibited [12,13] frequency bands or photonic band gaps. The DOS inside a photonic band gap (PBG) is automatically zero. It is proposed in literature [14-17] that these conditions might result in classical light localization, inhibition of single-photon emissions, fractionalized single-atom inversions, photon-atom bound states, and anomalously strong vacuum Rabi splitting.

There have been rigorous investigations of spontaneous decay of two-level atoms coupled with narrow cavity resonance according to Hermitian "universal" modes as against the dissipative quasi-modes of the cavity by reference [18]. These have concentrated in particular on cases in which the atomic line-width is nearly equal to the cavity resonance width γ, the so-called strong-coupling regime, when significant corrections are found into the golden rule. When the quality factor Q of cavity resonance falls within intermediate values, then spontaneous emission is seen to decay rapidly.

The emission processes investigated in this chapter regard a 1D unenclosed cavity, analysed according to the theory of reference [18]. There is specific discussion of the stimulated release of an atom under strong coupling regime inside a 1D-PBG cavity generated by two colliding laser beams. Atom-e.m. field coupling is modelled by quantum electro-dynamics, as per reference [18], with the atom considered as a two tier system, and the e.m. field as a superposition of normal modes. The coupling is in dipole approximation, Wigner-Weisskopf and rotating wave approximations are applied for the motion equations. An unenclosed cavity is conceived in the Quasi-Normal Mode (QNM), as in reference [18], and so the local density of states (LDOS) is defined as the local probability density of exciting a single cavity QNM. As a result the local DOS is effectively dependent on the phase difference between the two laser beams.

1.1. Quasi-Normal Modes (QNMs)

Describing a field inside an unenclosed cavity presents a problem that various authors have confronted, with references [19-22] proposing a QNM-based description of an electromagnetic field inside an open, one-sided homogeneous cavity. Because of the leakage, the QNMs exhibit complex eigen-frequencies as a consequence of leakage from the unenclosed cavity, with an orthogonal basis being assumed only inside the cavity and following a non-canonical metric.

The QNM approach was extended to open double-sided, non-homogeneous cavities and specifically to 1D-PBG cavities in references [23,24].

It is only possible to quantize a leaky cavity [25], considered a dissipative system, if the container is viewed as part of the total universe, within which energy is conserved [26]. A fundamental step towards the application of QNMs to the study of quantum electro-dynamics phenomena in cavities was already achieved in reference [27].

The second QNM-theory-based quantization scheme was extended to 1D-PBGs in reference [28]. References [29,30] applied the second QNM quantization to 1D-PBGs, excited by two pumps acting in opposite directions. The commutation relations observed for QNMs are not canonical, while also depending on the phase-difference between the two pumps and the unenclosed cavity geometry. Reference [31] applies QNM theory in an investigation of stimulated emission from an atom embedded inside a 1D-PBG under weak coupling regime, with two counter-propagating laser beams used to pump the system. The most significant result in reference [31] is the observation that the position of the dipole inside the cavity controls decay-time. This means that the phase-difference of the two laser beams can be used to control decay-time, which could be applicable on an atomic scale for a phase-sensitive, single-atom optical memory device.

The present chapter discusses stimulated emission of an atom enclosed inside a 1D-PBG cavity under strong coupling regime, generated by two counter-propagating laser beams [32,33]. The principal observation is a demonstration that high LDOS values can be used as a definition for a strong coupling regime. Further observations agree with literature in stating that the atomic emission probability decays with an oscillating pattern, and the atomic emission spectrum split into two peaks, known as Rabi splitting. What makes the observations of this chapter unique compared to literature is that by varying the laser beam phase difference it is possible to effectively control both the atomic emission probability oscillations, and the characteristic Rabi splitting of the emission spectrum. Some criteria are proposed for the design of active cavities, comprising a 1D-PBG together with atom, as active delay line, when it is possible to achieve high transmission in a narrow pass band for a delayed pulse by applying suitably differing laser beams phases.

In section 2, quantum electro-dynamic equations are used to model the coupling of an atom to an e.m. field, as an analogy of the theory of an atom in free space. In section 3, the atom is also contained within an unenclosed cavity, and the local probability density of a single QNM being excited is considered a definition of LDOS probability density. The atomic emission processes are modelled in section 4 with the LDOS of the stimulated emission depending on the phase difference of two counter-propagating laser beams. Section 5 discusses the probability of atomic emission under strong coupling regime. In section 6, the atomic emission spectrum is defined on the basis of its poles. Some criteria for the design of an active delay line are proposed in section 7, while section 8 is dedicated to a final discussion and concluding remarks.

2. Coupling of an atom to an electromagnetic (e.m.) field

The present case study examines an atom coupled to an e.m. field at a point $x_0 \in U$ inside a one-dimensional (1D) universe $U = \{x \mid x \in [-L/2, L/2], L \to \infty\}$ with refractive index n_0, and an unenclosed cavity $C = \{x \mid x \in (0, d), d < L\}$ with an inhomogeneous refractive index $n(x)$.

The atom is quantized into two levels, with an oscillating resonance of Ω [25]. The 1D universe modes are applied to quantize the e.m. field

$$\begin{cases} g_\lambda(x) = \dfrac{1}{\sqrt{L}} \exp\left(i\omega_\lambda \sqrt{\rho_0} x\right) \\ \omega_\lambda = \lambda \dfrac{\pi}{(L/2)\sqrt{\rho_0}} \quad , \quad \lambda \in \mathbb{Z} \end{cases}, \tag{1}$$

when $\rho_0 = (n_0/c)^2$, and c is the speed of light in a vacuum. The dipole operator μ [25] is used to model the atom along the direction of polarization of the e.m. field, and the coupling of the atom to the e.m. field is described using the electric dipole approximation [26].

At start time (t=0) the atom is in an excited state $|+\rangle$ and the e.m. field is in a vacuum state $|\{0\}\rangle = \prod_{\lambda=-\infty}^{\infty} |0_\lambda\rangle$. The system dynamics under these initial conditions can be described with basis states and corresponding eigen-values [18] as follows:

$$\begin{aligned} &|+,\{0\}\rangle = |+\rangle|\{0\}\rangle \quad , \quad \varepsilon_+ \\ &|-,1_\lambda\rangle = |-\rangle|1_\lambda\rangle \quad , \quad \varepsilon_- + \hbar\omega_\lambda \quad , \quad \lambda \in \mathbb{Z}. \end{aligned} \tag{2}$$

when $|+, \{0\}\rangle$ denotes the upper state of the atom, without any photons in all the e.m. modes; and $|-, 1_\lambda\rangle$ denotes the lower state of the atom, with one photon in the λth e.m. mode but any photons in the other e.m. modes.

2.1. Quantum electro-dynamic equations

If an initial condition $|+, \{0\}\rangle$ is assumed, then the atom-field system state at time instant t>0 can be defined as

$$|\psi(t)\rangle = c_+(t)|+,\{0\}\rangle + \sum_{\lambda=-\infty}^{\infty} c_{-,\lambda}(t)|-,1_\lambda\rangle, \tag{3}$$

introducing the probability amplitudes $c_+(t)$ and $c_{-,\lambda}(t)$ with $c_+(0)$=1 and $c_{-,\lambda}(0)$=0, and assuming the rotating wave approximation [26]. The time evolution equations for the probability amplitudes $c_+(t)$ and $c_{-,\lambda}(t)$ [18] are used as a starting point, thus

$$\begin{aligned} \frac{dc_+}{dt} &= \frac{\langle+|\mu|-\rangle}{\hbar}\sum_{\lambda=1}^{\infty}\sqrt{\frac{\hbar\omega_\lambda}{2\varepsilon_0 n_0^2}}\, g_\lambda(x_0)c_{-,\lambda}(t)\exp\left[-i(\omega_\lambda-\Omega)t\right] \\ \frac{dc_{-,\lambda}}{dt} &= -\frac{\langle-|\mu|+\rangle}{\hbar}\sqrt{\frac{\hbar\omega_\lambda}{2\varepsilon_0 n_0^2}}\, g_\lambda^*(x_0)c_+(t)\exp\left[i(\omega_\lambda-\Omega)t\right], \end{aligned} \tag{4}$$

the second of these can be formally integrated producing a time evolution equation for the probability amplitude $c_+(t)$ as follows:

$$\frac{dc_+}{dt} = -\frac{M}{2\varepsilon_0 n_0^2\hbar^2}\frac{1}{L}\sum_{\lambda=1}^{\infty}\hbar\omega_\lambda\int_0^t c_+(\tau)\exp\left[i(\omega_\lambda-\Omega)(\tau-t)\right]d\tau. \tag{5}$$

when ε_0 is the dielectric constant in vacuum and $M = |\langle+|\mu|-\rangle|^2$.

It is possible to establish a correspondence between the discrete modes and continuous modes of, respectively, a 1D cavity of length L, and an infinite universe of length $L \to \infty$. As $L \to \infty$, the mode spectrum approaches continuity, since $\Delta\omega_\lambda = \pi / (L/2)\sqrt{\rho_0} \approx d\omega \to 0$. When this limit is reached, sums over discrete indices can be transformed into integrals over a continuous variable of frequency,

$$\frac{1}{L}\sum_{\lambda=1}^{\infty} \Rightarrow \int_0^{\infty} d\omega \sigma^{(\mathrm{loc})}(x_0,\omega), \tag{6}$$

when $\sigma^{(\mathrm{loc})}(x,\omega)$ is the local density of states (LDOS), which can be interpreted as the density of probability for an excited level of the e.m. field, at a point x, collapsed into a single eigen-state, oscillating around the frequency ω [34,35]. Strictly speaking, in Equation (6), the range of integration over ω only extends from 0 to ∞, given that the physical frequencies are defined as positive. Nevertheless, it is possible to extend the range from −∞ to ∞ without significant errors, due to the fact that most optical experiments utilize a narrow band source B [36], such that $B \ll \omega_{c,}$ with ω_c the bandwidth B as the central frequency. The time evolution Equation (5) therefore becomes

$$\frac{dc_+}{dt} = \int_0^t d\tau \cdot K(x_0, t-\tau) c_+(\tau), \tag{7}$$

the kernel function $K(x, t)$ being defined as:

$$K(x,t) = -\frac{M}{2\varepsilon_0 n_0^2 \hbar^2} \int_{-\infty}^{\infty} d\omega \hbar\omega \sigma^{(\mathrm{loc})}(x,\omega) \exp\left[-i(\omega-\Omega)t\right]. \tag{8}$$

As emerges from Equation (8), there is a marked dependence of the kernel function on the LDOS through $\sigma^{(\mathrm{loc})}(x,\omega)$. It is possible to reinterpret the latter as the density of photon states in the reservoir. Essentially, the kernel function (8) is a gauge of the memory of previous state of the photon reservoir, within the evolutionary time scale of the atomic system, thus $K(x, t)$ could be considered the photon reservoir's memory kernel.

2.2. Atom in free space

If an atom is located at a given point x_0 outside the unenclosed cavity, so that $x_0<0$ or $x_0>d$, then the local DOS $\sigma^{(\mathrm{loc})}(x,\omega)$ refers to free space (see references [26,31]):

$$\sigma^{(\mathrm{loc})}(x,\omega) = \sigma_{\text{free-space}}(\omega) = \sigma_{\mathrm{ref}} = \frac{\sqrt{\rho_0}}{2\pi}. \tag{9}$$

The probability of atomic emission decays exponentially in free space,

$$|c_+(t)|^2 = \exp(-\Gamma_0 t) \quad , \quad t \geq 0, \tag{10}$$

being Γ_0 the atomic decay rate:

$$\Gamma_0 = \frac{M}{\hbar}\frac{\sqrt{\rho_0}}{\varepsilon_0 n_0^2}\Omega \Bigg/ \left[1 + \frac{1}{4}\left(\frac{M}{\hbar}\frac{\sqrt{\rho_0}}{\varepsilon_0 n_0^2}\right)^2\right]. \tag{11}$$

Free space is an infinitely large photon reservoir (a flat spectrum), and so it should respond instantaneous, with any memory effects associated to emission dynamics being infinitesimally short relative to any time intervals of interest. According to the so-called Markovian [26] interactions, an excited state population gradually decays to ground level in free space, regardless of any driving field strength. This result is generally valid for almost any smoothly varying broadband DOS.

The following parameter is now introduced as a step for the analysis of the next section,

$$R = \frac{\Gamma_0}{\Omega} \cong \frac{M}{\hbar}\frac{\sqrt{\rho_0}}{\varepsilon_0 n_0^2}, \tag{12}$$

interpretable as the degree of atom-field coupling, and with the possibility of expressing Equation (8) as:

$$K(x,t) = -\frac{R}{2\sqrt{\rho_0}}\int_{-\infty}^{\infty} d\omega \cdot \omega\sigma^{(\mathrm{loc})}(x,\omega)\exp\left[-i(\omega-\Omega)t\right]. \tag{13}$$

3. Atom inside an unenclosed cavity

Assuming $0<x_0<d$, which represents an atom embedded at a point x_0 inside an open, inhomogeneous cavity with refractive index $n(x)$, if the resonance frequency Ω of the atom is coupled with the nth QNM oscillating at frequency $\mathrm{Re}\omega_n$, then the coupling will exhibit frequency detuning:

$$\Delta_n = \frac{\mathrm{Re}\,\omega_n - \Omega}{R}. \tag{14}$$

3.1. Density Of States (DOS) as the probability density to excite a singleQNM

By filtering two counter-propagating pumps at an atomic resonance $\Omega \approx \mathrm{Re}\omega_n$, it emerges that only the nth QNM, and no other QNMs, can be excited, because the nth QNM is the only one oscillating at the frequency $\mathrm{Re}\omega_n$ and within the narrow range $2|\mathrm{Im}\omega_n| << |\mathrm{Re}\omega_n|$, which is sufficiently remote to exclude the other QNMs [23,24]. Around point x, the local probability density that the e.m. field is in fact excited on the nth QNM is [31]

$$\sigma_n^{(\mathrm{loc})}(x,\omega) = \frac{1}{I_n}\sigma_n(\omega)\rho(x)\left|f_n^{\mathrm{N}}(x)\right|^2 \quad , \quad \forall x \middle| 0 < x < d, \tag{15}$$

which is related directly to the (integral) probability density $\sigma_n(\omega)$ for the nth QNM. In Equation (15), $\rho(x)=[n(x)/c]^2$, I_n denotes an appropriate overlapping integral [28], while $f_n^{\mathrm{N}}(x)=f_n(x)\sqrt{2\omega_n/\langle f_n | f_n\rangle}$ is the normalized QNM function, with $\langle f_n | f_n\rangle$ representing the QNM norm.

For the investigation of spontaneous emissions, the two pumps are modelled as fluctuations of vacuum, based on the e.m. field ground state (for examples, see references [26,28-30]). The (integral) probability density that the nth QNM is excited within the unenclosed cavity can be expressed as [31]:

$$\sigma_n^{(\mathrm{I})}(\omega) = \frac{1}{d}\int_0^d \sigma_n^{(\mathrm{I})}(x,\omega)dx = \alpha_n \frac{d}{2\pi} \frac{I_n^2|\mathrm{Im}\,\omega_n|}{(\omega - \mathrm{Re}\,\omega_n)^2 + \mathrm{Im}^2\,\omega_n}. \tag{16}$$

It is possible to deduce a normalization constant α_n from the condition:

$$\int_{\mathrm{Re}\,\omega_n - |\mathrm{Im}\,\omega_n|}^{\mathrm{Re}\,\omega_n + |\mathrm{Im}\,\omega_n|} \sigma_n^{(\mathrm{I})}(\omega)d\omega + \int_{-\mathrm{Re}\,\omega_n - |\mathrm{Im}\,\omega_n|}^{-\mathrm{Re}\,\omega_n + |\mathrm{Im}\,\omega_n|} \sigma_n^{(\mathrm{I})}(\omega)d\omega = \frac{1}{d}. \tag{17}$$

From Equation (16) it was deduced that the probability density due to fluctuations in vacuum for the nth QNM is a Lorentzian function, with parameters including real and imaginary parts of the nth QNM frequency. There is a relation between the overlapping integral I_n and the statistical weight of the nth QNM in the DOS. Equation (17) also integrates the probability density $\sigma_n^{(\mathrm{I})}(\omega)$ into the range of negative frequencies $\omega \in [-\mathrm{Re}\omega_n - |\mathrm{Im}\omega_n|, -\mathrm{Re}\omega_n + |\mathrm{Im}\omega_n|]$, since with $\mathrm{Re}\omega_n>0$ the QNM frequency ω_n is also represented by frequency $\omega_{-n}=-\omega_n^*$ with $\mathrm{Re}\omega_{-n}<0$ [23,24].

Stimulated emissions are considered by modelling the two pumps as a pair of laser beams in a coherent state (see references [26,28-30] for examples). When the symmetry property is achieved by the refractive index $n(x)$, so that $n(d/2-x)=n(d/2+x)$, then the probability density that the e.m. field is excited to the nth QNM inside the cavity can be written as [31]:

$$\sigma_n^{(\mathrm{II})}(\omega) = \sigma_n^{(\mathrm{I})}(\omega)\left[\Delta + (-1)^n \cos\ \varphi\right] \tag{18}$$

Equation (18) shows that the phase-difference $\Delta\varphi$ of the pair of counter-propagating laser beams can be used to control the probability density for the nth QNM.

4. Atomic emission processes

With an atom at point x_0 of an unenclosed inhomogeneous cavity, so that $0<x_0<d$, and electromagnetic field coupling limited to the nth QNM $[\omega_n, f_n^N(x)]$ of the unenclosed cavity, then the local probability density $\sigma_n^{(loc)}(x,\omega)$ is related to the integral probability density $\sigma_n(\omega)$ as in Equation (15) and the atomic emission processes exhibit a characteristic kernel function $K(x,t)$, which can be expanded as [see Equation (13)]:

$$K_n(x,t) = -\frac{R}{2\sqrt{\rho_0}}\frac{1}{I_n}\rho(x)\left|f_n^N(x)\right|^2 \int_{-\infty}^{\infty} d\omega\cdot\omega\sigma_n(\omega)\exp\left[-i(\omega-\Omega)t\right] \quad , \quad \forall x|0<x<d. \tag{19}$$

4.1. Spontaneous emission: DOS due to vacuum fluctuations

If the unenclosed cavity is only affected by fluctuations of vacuum filtered at the atomic resonance Ω and close to the cavity's n^{th} QNM frequency ($\Omega\approx\mathrm{Re}\omega_n$), then Equations (16) and (17) can be used to express the integral probability density $\sigma_n^{(I)}(\omega)$ for the nth QNM, and atomic spontaneous emission exhibits a characteristic time evolution [see Equation (7)] in which the kernel function $K_n(x,t)$ can be expressed as [see Equation (19)]:

$$\begin{aligned} K_n^{(I)}(x,t) = &-\frac{R}{2\sqrt{\rho_0}}\frac{1}{I_n}\rho(x)\left|f_n^N(x)\right|^2 \alpha_n \frac{d}{2\pi} I_n^2 \exp(i\Omega t)\cdot \\ &\cdot i\sqrt{2\pi}\left[\frac{1}{\sqrt{2\pi}}\int_{-\infty}^{\infty}(-i\omega)\frac{|\mathrm{Im}\,\omega_n|}{(\omega-\mathrm{Re}\,\omega_n)^2+\mathrm{Im}^2\,\omega_n}\exp(-i\omega t)\,d\omega\right]. \end{aligned} \tag{20}$$

The resulting signal $x(t)$ can easily be transformed into the Fourier domain [37]:

$$\begin{aligned} &x(t) = \frac{1}{2}\frac{d}{dt}\left[e^{-(|\mathrm{Im}\,\omega_n|+i\,\mathrm{Re}\,\omega_n)|t|}\right] \\ &\Rightarrow X(\omega) = \frac{1}{\sqrt{2\pi}}\int_{-\infty}^{\infty} x(t)\exp(i\omega t)dt = -\frac{i\omega}{\sqrt{2\pi}}\frac{|\mathrm{Im}\,\omega_n|}{\mathrm{Im}^2\,\omega_n+(\omega-\mathrm{Re}\,\omega_n)^2}. \end{aligned} \tag{21}$$

Applying Equations (20) and (21) for the kernel function of the spontaneous emission process gives:

$$K_n^{(I)}(x,t) = -\frac{1}{4}\frac{R}{\sqrt{\rho_0}}\alpha_n(d\cdot I_n)\omega_n\rho(x)\left|f_n^N(x)\right|^2 \exp\left[-i(\omega_n-\Omega)t\right] \quad , \quad \forall t\geq 0. \tag{22}$$

Now, in Equation (7), deriving under the integral sign [37] gives

$$\frac{d^2c_+}{dt^2} = K_n^{(\mathrm{I})}(x_0,t=0)c_+(t) + \int_0^t d\tau \frac{\partial K_n^{(\mathrm{I})}}{\partial t} c_+(\tau), \tag{23}$$

and deriving Equation (22) again, sampled at point x_0, relative to time,

$$\frac{\partial K_n^{(\mathrm{I})}}{\partial t} = -i(\omega_n - \Omega)K_n^{(\mathrm{I})}(x_0,t) \quad , \quad t \geq 0, \tag{24}$$

which after some algebraic transformations produces a second order differential equation in time for spontaneous emission probability:

$$\begin{aligned} &\frac{d^2c_+}{dt^2} + i(\omega_n - \Omega)\frac{dc_+}{dt} - K_n^{(\mathrm{I})}(x_0,t=0)c_+(t) = 0 \\ &c_+(0) = 1 \quad , \quad \left.\frac{dc_+}{dt}\right|_{t=0} \propto c_{-,n}(0) = 0. \end{aligned} \tag{25}$$

4.2. Stimulated emission: DOS dependant on the phase difference of a pair of counter-propagating laser-beams

If the unenclosed cavity is pumped coherently by two counter-propagating laser beams with a phase difference $\Delta\varphi$, tuned to the atomic resonance Ω and closed to the nth QNM frequency ($\Omega \approx \mathrm{Re}\omega_n$), then the probability density $\sigma_n^{(\mathrm{II})}(\omega)$ for the nth QNM, for the state of the two laser beams, is related to $\sigma_n^{(\mathrm{I})}(\omega)$, which is calculated using Equation (18) when vacuum fluctuations are present. The atomic stimulated emission exhibits a characteristic kernel function $K_n(x,\omega)$, which can be expressed as follows [see Equation (20)]:

$$K_n^{(\mathrm{II})}(x,t) = K_n^{(\mathrm{I})}(x,t)\left[1 + (-1)^n \cos\Delta\varphi\right]. \tag{26}$$

The quantity $(\omega_n - \Omega)$ can be re-expressed in terms of frequency detuning (14), as $(\omega_n - \Omega) = (\mathrm{Re}\omega_n - \Omega) + i\mathrm{Im}\omega_n = R\Delta_n + i\mathrm{Im}\omega_n$ so that the second order differential equation for emission probability becomes:

$$\frac{d^2c_+}{dt^2} + i(R\Delta_n + i\,\mathrm{Im}\,\omega_n)\frac{dc_+}{dt} - K_n(x_0,t=0)c_+(t) = 0 \tag{27}$$

5. Atomic emission probability

The initial conditions being the same as Equation (25), the algebraic equation associated with the Cauchy problem (27) can be recast as:

$$p^2 + \left(R\Delta_n + i\,\mathrm{Im}\,\omega_n\right)p + K_n\left(x_0, t = 0\right) = 0. \tag{28}$$

This is solved with two roots,

$$p_{1,2} = \frac{\left(R\Delta_n + i\,\mathrm{Im}\,\omega_n\right)}{2}\left[-1 \pm \sqrt{1 - \frac{4K_n\left(x_0, t = 0\right)}{\left(R\Delta_n + i\,\mathrm{Im}\,\omega_n\right)^2}}\right], \tag{29}$$

which permit expression of the particular integral of the differential Equation (27) as:

$$c_+\left(t\right) = \frac{p_2}{p_2 - p_1}\exp\left(ip_1 t\right) - \frac{p_1}{p_2 - p_1}\exp\left(ip_2 t\right). \tag{30}$$

The atom and the nth QNM are coupled under a strong regime when the behaviour of the particular integral (30) is oscillatory, and when the two roots (29) of the relative algebraic Equation (28), are complex conjugates [18].

5.1. Strong coupling regime

Spontaneous emission is examined in order to assess the atom - n^{th} QNM coupling under strong regime. Given that,

$$4\left|\frac{K_n^{(1)}\left(x_0, t = 0\right)}{\left(R\Delta_n + i\,\mathrm{Im}\,\omega_n\right)^2}\right| < \frac{R}{d \cdot I_n}\frac{\sigma_n^{(1)}\left(x_0, \Omega\right)}{\sigma_{\text{free-space}}} \cong R\frac{\sigma_n^{(1)}\left(x_0, \Omega\right)}{\sigma_{\text{free-space}}}, \tag{31}$$

there is a strong coupling regime if [18]:

$$4\left|\frac{K_n\left(x_0, t = 0\right)}{\left(R\Delta_n + i\,\mathrm{Im}\,\omega_n\right)^2}\right| > 1 \;\Rightarrow\; \frac{\sigma_n^{(1)}\left(x_0, \Omega\right)}{\sigma_{\text{free-space}}} > \frac{1}{R}. \tag{32}$$

Equation (32) shows that a strong coupling regime is present when the probability density (16) inside the unenclosed cavity, sampled at atomic resonance in units of DOS (9) with reference to free space, is in excess of the inverse of the atomic parameter R [see Equation (12)]. An interpretation of parameter R as a level of atom field coupling is thus legitimated: the greater R becomes, the better Equation (32) is satisfied. The two roots (29) become complex conjugates in the hypothesis of a strong coupling regime (32),

$$p_{1,2} \cong -\frac{R\Delta_n + i\,\mathrm{Im}\,\omega_n}{2} \pm i\sqrt{K_n^{(1)}\left(x_0, t = 0\right)}, \tag{33}$$

and the behaviour of the particular integral (30) is oscillatory:

$$c_+(t) \cong \exp\left(-i\frac{R\Delta_n + i\,\mathrm{Im}\,\omega_n}{2}t\right)\cdot \\ \cdot\left\{\cosh\left[\sqrt{K_n^{(\mathrm{I})}(x_0,t=0)}t\right] + i\frac{R\Delta_n + i\,\mathrm{Im}\,\omega_n}{2\sqrt{K_n^{(\mathrm{I})}(x_0,t=0)}}\sinh\left[\sqrt{K_n^{(\mathrm{I})}(x_0,t=0)}t\right]\right\}. \tag{34}$$

In reality [see Equation (22)] $K_n^{(\mathrm{I})}(x_0, t=0) = -K_n^{(\mathrm{I})}(x_0, t=0)$. It is possible to interpret the oscillatory behaviour as emission re-absorption of a single photon and so the net decay rate can thus be determined from the rate of photon leakage, which is $|\mathrm{Im}\omega_n|/2$.

In the case of stimulated emissions, the coupling between atom and the nth QNM can again be considered under strong regime. Given a phase difference of $\Delta\varphi$ for the pair of counter-propagating laser beams, then the atom - n^{th} QNM coupling exhibits the kernel function (26). Assuming hypothetical strong coupling as expressed by a similar condition to Equation (32), the behaviour of the particular integral (30) is oscillatory,

$$c_+(t) \cong \exp\left(-i\frac{R\Delta_n + i\,\mathrm{Im}\,\omega_n}{2}t\right)\cdot \\ \cdot\left\{\cosh\left[\sqrt{K_n^{(\mathrm{II})}(x_0,t=0)}t\right] + i\frac{R\Delta_n + i\,\mathrm{Im}\,\omega_n}{2\sqrt{K_n^{(\mathrm{II})}(x_0,t=0)}}\sinh\left[\sqrt{K_n^{(\mathrm{II})}(x_0,t=0)}t\right]\right\}, \tag{35}$$

when $K_n^{(\mathrm{II})}(x_0, t=0)$ is linked to the phase difference $\Delta\varphi$ through Equation (26). The quantity of atomic emission probability oscillations is dependent on the position of the atom inside the cavity and so the phase-difference of the paired laser-beams can be used to control it. The condition,

$$1+(-1)^n\cos\Delta\varphi = 0, \tag{36}$$

is satisfied if the atom is coupled to an odd QNM, i.e. n=1,3,... and the paired laser beams are in phase, i.e. $\Delta\varphi$=0; or if the atom is coupled to an even QNM, i.e. n=0,2,... and the paired laser beams are out of phase, i.e. $\Delta\varphi=\pi$. When Equation (36) is satisfied, the probability of emission is over-damped within the entire cavity. Even under strong coupling, no oscillation occurs [see Equation (35)]:

$$K_n^{(\mathrm{II})}(x_0,t=0)=0 \;\Rightarrow\; |b_+(t)|^2 = \exp\left(\frac{|\mathrm{Im}\,\omega_n|^2}{4}t\right)\left[\left(1+\frac{|\mathrm{Im}\,\omega_n|}{2}t\right)^2 + \left(\frac{R\Delta_n}{2}t\right)^2\right]. \tag{37}$$

6. Atomic emission spectrum

An atom located at point x_0 is in its upper state at initial time (t=0) and there are no photons present in any normal mode, i.e. $c_+(x_0, t=0)=1$. Following atomic decay ($t=\infty$), Equation (4) can be used to derive the coefficient of probability $c_{-,\lambda}(x_0, t)$ of finding the atom in its lower state with one photon in the λth e.m. mode and no photons in all the other modes:

$$c_{-,\lambda}(x_0, t=\infty) = -\frac{\langle -|\mu|+\rangle}{\hbar}\sqrt{\frac{\hbar\omega_\lambda}{2\varepsilon_0 n_0^2}}g_\lambda^*(x_0)\int_0^\infty c_+(x_0, t)\exp\left[i(\omega_\lambda - \Omega)t\right]dt. \tag{38}$$

Applying the Laplace transform for probability coefficient $c_+(x_0, t)$,

$$C_+(x_0, s) = \frac{1}{\sqrt{2\pi}}\int_0^\infty c_+(x_0, t)\exp(-st)dt, \tag{39}$$

Equation (38) can thus be re-formulated as:

$$c_{-,\lambda}(x_0, t=\infty) = -\frac{\langle -|\mu|+\rangle}{\hbar}\sqrt{\frac{\hbar\omega_\lambda}{2\varepsilon_0 n_0^2}}g_\lambda^*(x_0)\sqrt{2\pi}C_+\left[x_0, s=i(\omega_\lambda - \Omega)\right]. \tag{40}$$

If decay has occurred ($t=\infty$), it is possible to define the atomic emission spectrum as the probability density that the atom at point x_0 emitted at frequency ω [18], i.e.

$$W(x_0, \omega) = \sum_{\lambda=1}^{\infty}\left|c_{-,\lambda}(x_0, t=\infty)\right|^2 \delta(\omega - \omega_\lambda), \tag{41}$$

when $\delta(t)$ is the Dirac delta distribution. Integrating Equation (40) into Equation (41), gives:

$$W(x_0, \omega) = \frac{M}{2\varepsilon_0 n_0^2 \hbar^2}\frac{1}{L}\sum_{\lambda=1}^{\infty}\hbar\omega_\lambda 2\pi\left|C_+\left[x_0, i(\omega_\lambda - \Omega)\right]\right|^2 \delta(\omega - \omega_\lambda). \tag{42}$$

If sums over discrete quantities are converted to integrals over continuous frequencies, using Equation (6), then the Dirac delta properties can be used to reduce the emission spectrum (42) to:

$$W(x_0, \omega) = \alpha' \cdot \frac{R}{2\sqrt{\rho_0}}\omega\sigma^{(\text{loc})}(x_0, \omega)2\pi\left|C_+\left[x_0, i(\omega - \Omega)\right]\right|^2, \tag{43}$$

when α' is a suitable normalization constant and $\sigma^{(loc)}(x,\omega)$ is the local density of states (DOS). Equation (12) is used to define the atomic parameter R.

Given that most optical experiments apply a narrow band source [36], it is possible to extend the frequency range from $-\infty$ to ∞ without significant errors, and the closure relation can be applied to establish the normalization constant α'

$$\int_{-\infty}^{\infty} W(x_0,\omega)d\omega = 1, \tag{44}$$

which derives directly from the interpretation of emission spectrum probability (43). Assuming $0<x_0<d$ and $n(x_0)>n_0$, the atom is embedded inside an unenclosed cavity with inhomogeneous refractive index $\rho(x)=[n(x)/c]^2$. The atom with resonance frequency Ω can be assumed coupled to the nth QNM and oscillating at the frequency $\mathrm{Re}\omega_n$. This atom to n^{th} QNM coupling is characterized by frequency detuning Δ_n (14). The normalization condition (44) can be reduced to:

$$2\int_{\Omega-|\mathrm{Im}\,\omega_n|}^{\Omega+|\mathrm{Im}\,\omega_n|} W_n(x_0,\omega)d\omega = 1. \tag{45}$$

Integrals over positive frequencies are multiplied by a factor of 2 in Equation (45) in order to include the contribution of negative frequencies [see comments following Equation (17)].

Equation (15) showed that the local probability density $\sigma_n^{(loc)}(x,\omega)$ for the nth QNM was proportional to $\sigma_n(\omega)$. Now if Equation (15) is included into Equation (43), the atomic emission spectrum is expressed as

$$W_n(x_0,\omega) = \alpha_n' \cdot \frac{R}{2\sqrt{\rho_0}}\frac{1}{I_n}\rho(x_0)\left|f_n^{(N)}(x_0)\right|^2 \omega\sigma_n(\omega)2\pi\left|C_+^{(n)}\left[x_0,i(\omega-\Omega)\right]\right|^2, \tag{46}$$

when α'_n is the normalization constant that satisfies Equation (45).

The atomic emission processes exhibit a characteristic kernel function $K_n(x,t)$, here expressible as in Equation (22), while for stimulated emission as in Equation (26). By including Equation (22) into Equation (46), the emission spectrum (46) assumes the form:

$$\frac{W_n(x_0,\omega)}{\alpha_n'} = -\frac{2}{d\cdot I_n^2}\frac{K_n^{(I)}(x_0,t=0)}{\omega_n}\omega\frac{\sigma_n(\omega)}{K_n}2\pi\left|C_+^{(n)}\left[x_0,i(\omega-\Omega)\right]\right|^2. \tag{47}$$

The emission spectrum $W_n(x_0,\omega)$ emerging from Equation (47) depends on both the probability density $\sigma_n(\omega)$, and the initial kernel function value $K_n^{(I)}(x_0, t=0)$.

Now, by applying the Laplace transformation of the Cauchy problem (27), and the initial conditions derived from Equation (25), gives finally [37]

$$C_{+}^{(n)}\left(x_0,i\xi\right)=-\frac{i}{\sqrt{2\pi}}\frac{\xi+R\Delta_n+i\,\mathrm{Im}\,\omega_n}{\xi^2+\left(R\Delta_n+i\,\mathrm{Im}\,\omega_n\right)\xi+K_n\left(x_0,t=0\right)},\tag{48}$$

when ξ is the shifted frequency (ω–Ω), with Ω denoting atomic resonance.

6.1. Poles of the emission spectrum

The two poles that solve Equation (28), p_1 and p_2, can be used to describe the atomic emission spectrum, as expressed in Equation (29). Hypothesizing a strong coupling regime [see Equation (32)], the atomic emission spectrum $W_n(x_0,\xi)$ as a function of the shifted frequency ξ =(ω–Ω), exhibits two characteristic peaks, centred approximately in the Rep_1 and Rep_2 resonances with bandwidths linked to 2 | Imp_1 | and 2 | Imp_2 |. There is thus a Rabi splitting with the two peaks separated by:

$$\Delta\xi=\mathrm{Re}\,p_1-\mathrm{Re}\,p_2.\tag{49}$$

Considering stimulated emission processes, the paired counter-propagating laser beams are set to a phase difference $\Delta\varphi$, and so the emission spectrum $W_n(x_0,\xi)$ can be described using a kernel function $K_n(x_0,t=0)$ associated with $\Delta\varphi$ [see Equation (26)]. The Rabi splitting thus depends not only on the position of the atom inside the cavity, but can also be imposed by the phase-difference of the paired laser-beams.

If the operative condition is close to that defined by Equation (36), such that $K_n(x_0,t=0)\approx 0$, the spectrum $W_n(x_0,\omega)$ as a function of the pure frequency ω is limited to two pulses that almost superimpose each other: 1) a Lorentzian function centred in the nth QNM frequency Reω_n, with bandwidth 2 | Imω_n |, superimposed on 2) a Dirac distribution of atomic resonance $\Omega\approx$Reω_n, so that [see Equations (16)-(18)]

$$W_n\left(x_0,\omega\right)\approx\sigma_n^{(\mathrm{II})}\left(\omega\right)+\frac{\alpha''_n}{d}\delta\left(\omega-\Omega\right)\rightarrow\delta\left(\omega-\Omega\right),\tag{50}$$

when α''_n is the normalization constant that satisfies condition (45). The two poles, ω_1 and ω_2, can be simplified as [see Equations (14), (28) and (29)]:

$$\begin{aligned}\omega_1&\approx\mathrm{Re}\,\omega_n+i\,\mathrm{Im}\,\omega_n\\ \omega_2&\approx\Omega.\end{aligned}\tag{51}$$

7. Criteria for designing an active delay line

In references [23,24] and subsequent papers, the QNM theory was applied to a photonic crystal (PC) as a symmetric Quarter-Wave (QW) 1D-PBG cavity. The present study considers a symmetric QW 1D-PBG cavity with parameters $\lambda_{ref}=1\mu m$, $N=5$, $n_h=2$, $n_l=1.5$ (see Figure 1). The motivation for choosing this cavity is that it provides a relatively simple physical context for discussion of criteria in order to design an active delay line. An atom is located in the centre of the 1D-PBG, so that $x_0=d/2$ (see Figure 1). Reference [28] discusses how in a symmetric QW 1D-PBG cavity with reference wavelength ω_{ref} and N periods, the $[0, 2\omega_{ref})$ range includes $2N+1$ QNMs, which are identified as $|n\rangle, n\in[0, 2N]$ (with the exclusion of $\omega=2\omega_{ref}$). If the location of the atom is the centre x_0 of the 1D-PBG cavity, then it can only be coupled to one of the even QNMs n because the QNM intensity $|f_n|^2$ in this position has a maximum for even values of n and is almost null for odd values of n.

The active cavity consists of the 1D-PBG cavity containing one atom, and it is characterized by a $G(x_0,\omega)$ global transmission spectrum, this being the product of the 1D-PBG $|t(\omega)|^2$ transmission spectrum, and the $W(x_0,\omega)$ emission spectrum of the atom [in units of s], so

$$G(x_0,\omega)=W(x_0,\omega)|t(\omega)|^2\left[\text{in units of } s\right]. \tag{52}$$

It is possible to define the active cavity's "density of coupling" (DOC) $\sigma_C(x_0,\omega)$, as the probability density that an atom embedded at point x_0, is coupled to only a single QNM, with a oscillation close to the frequency ω. The DOC $\sigma_C(x_0,\omega)$ [in units of s^2/m] is the product of the atomic emission spectrum $W(x_0,\omega)$, and the DOS $\sigma(\omega)$ [in units of s/m]. It is possible to introduce an "acceleration of coupling" $a_C(x_0, \omega)$ inside the active cavity as:

$$a_C(x_0,\omega)=\frac{1}{\sigma_C(x_0,\omega)}=\frac{1}{W(x_0,\omega)\sigma(\omega)}=\frac{\mathrm{v}(\omega)}{W(x_0,\omega)}\left[\text{in units of } m/s^2\right]. \tag{53}$$

when $\mathrm{v}(\omega)=1/\sigma(\omega)$. If the active cavity is to be designed as an ideal delay line, then the pulsed input needs to be retarded and highly amplified, but free of any distortion. For a narrow pass band the global transmission (52) needs to be very high, with a quasi-constant acceleration of coupling (53).

As described above, an atom embedded in the centre of a symmetric QW 1D-PBG cavity with $N=5$ periods (Figure 1) can only be coupled to a single QNM, oscillating close to an even transmission peak $n=0,2,...,2N$. If it is assumed that the atom is coupled to the $(N+1)^{th}$ QNM, close to the edge of the high frequency band, then the 1D-PBG cavity quality factor will be

$$Q_{N+1}=\frac{\Omega}{|\mathrm{Im}\,\omega_{N+1}|}, \tag{54}$$

when Ω is the atom's resonance frequency. If it is also assumed that a strong coupling regime is in force, then the active delay line directories can be satisfied by an appropriate coupling degree value,

$$R = \frac{\Gamma_0}{\Omega}, \tag{55}$$

when Γ_0 denotes the atomic decay rate in vacuum, and by a suitable atomic frequency detuning value - $(N+1)^{th}$ QNM coupling,

$$\Delta_{N+1} = \frac{\mathrm{Re}\,\omega_{N+1} - \Omega}{R}. \tag{56}$$

If spontaneous emission occurs, assuming perfect tuning so that $\Delta_{N+1}=0$, then the oscillation of the atom is at the frequency of the $(N+1)^{th}$ QNM, which is $\Omega=\mathrm{Re}\omega_{N+1}$. A suitable value of coupling degree R thus exists (see Figures 2.a and 2.b) as $R^*=0.002506$, making the two poles [Equations (29) and (22)] of the atomic emission spectrum distinct for $R>R^*$ or coincident for $0<R<R^*$. Alternatively stated, when $R>R^*$, there is a Rabi splitting (see Figure 3.a) in the atomic emission spectrum [Equations (47), (48) and (22)], generating an oscillation (see Figure 4.a) in the atomic emission probability [Equations (34) and (22)]. Conversely, when $0<R<R^*$, the emission spectrum comprises two superimposed peaks, indicating over-damping of the emission probability. In an attempt to identify Rabi splitting under strong coupling and consistent with experimentation ($\Gamma_0 \sim |\mathrm{Im}\omega_{N+1}|$) [38], the following degree of coupling is postulated:

$$R = R_{N+1} = \frac{1}{Q_{N+1}}. \tag{57}$$

The two spontaneous emission spectrum poles, shifted by the atomic resonance Ω, are $\xi_1=0.06383+i0.01770$ and $\xi_2=-0.06383+i0.01995$ in units of ω_{ref} (see Figures 2.a and 2.b). They describe the two emission spectrum peaks in resonance and bandwidth, with maxima of $W_1=21.87$ and $W_2=15.66$ in units of ω_{ref} (see Figure 3.a). Assuming the disappearance of emission probability after the second oscillation, then the decay time value is $\tau=94.3$ in units of $1/\omega_{\mathrm{ref}}$ (see Figure 4.a). The active cavity designed in this way is a less than ideal optical amplifier, in the sense that the amplification of an input pulse is accompanied by distortion. In the case of spontaneous emission plotted in Figures 5.a and 5.b the pass band is narrow, with $\xi = \omega-\Omega \approx (-0.06, 0.06)$ (in units of ω_{ref}), where the global transmission spectrum exhibits relatively high values, $G_{C,N+1} \in (G_{min}, G_{max})=(2.881,\ 14.43)$ in units of $1/\omega_{\mathrm{ref}}$. While the coupling acceleration is modulated close to the value $v_{C,N+1}=0.03445$ in units of $\omega_{\mathrm{ref}}/v_{\mathrm{ref}}$.

An example of stimulated emission is now considered, with the atom inside the symmetric QW 1D-PBG cavity being excited by a pair of counter-propagating laser beams. The phase difference $\Delta\varphi$ of the two laser beams can thus be added as a new degree of freedom for the realization of an active delay line. If perfect tuning is assumed during stimulated emission, when $\Delta_{N+1}=0$, the atom again oscillates at the frequency of the $(N+1)^{th}$ QNM, which is $\Omega=\mathrm{Re}\omega_{N+1}$. The phase difference range $\Delta\varphi$, this being $(\Delta\varphi_1, \Delta\varphi_2)=(2.747, 3.524)$ in *rad* units, is adequate (see Figures 2.c and 2.d) to make the two poles [Equations (29) and (26)] of the atomic emission spectrum distinct for $\Delta\varphi<\Delta\varphi_1$ and $\Delta\varphi>\Delta\varphi_2$, but coincident for $\Delta\varphi_1<\Delta\varphi<\Delta\varphi_2$. In other terms, when $\Delta\varphi<\Delta\varphi_1$ and $\Delta\varphi>\Delta\varphi_2$ Rabi splitting (see Figure 3.a) occurs in the atomic emission spectrum [Equations (47), (48) and (26)], with an oscillation (see Figure 4.a) in the probability of atomic emission [Equations (35) and (26)]. When $\Delta\varphi_1<\Delta\varphi<\Delta\varphi_2$, the emission spectrum comprises two superimposed peaks with over-damping of emission probability. The Rabi splitting and decay time oscillations can therefore be controlled using the phase difference of the paired laser beams.

Using stimulated emission to obtain an ideal delay line, requires that the paired laser beams have higher quadrature, so $\Delta\varphi>\pi/2$. Compared to spontaneous emission, the emission spectrum must show narrower Rabi splitting and the emission probability must have a longer decay time. The active cavity comprising the 1D-PBG together with the atom can thus act as a delay line, because the active cavity delay time is linked to the atomic decay time (for examples, see references [39-41]). As noted above, then it is necessary that the phase difference remains within a maximum of $\Delta\varphi_1=2.747$ (in *rad* units), beyond which the Rabi splitting tends towards zero. In the same time domain, increasing the phase difference relative to $\Delta\varphi \approx \pi/2$, causes the decay time to become even longer, while in the frequency domain the global transmission (52) exhibits high gain but instead the acceleration of coupling (53) exhibits a narrow pass band. The active cavity thus acts as an active but not ideal delay line when $\Delta\varphi \rightarrow \Delta\varphi_1$. This leads to the conclusion that the 1D-PBG cavity should be pumped by paired laser beams exceeding a tilt angle quadrature of:

$$\Delta\varphi = \frac{\pi}{2} + \frac{\pi}{10}. \tag{58}$$

The two stimulated emission spectrum poles are shifted by the resonance Ω, and are respectively $\xi_1=0.05205+i0.01787$ and $\xi_2=-0.05205+i0.01978$ (in units of ω_{ref}) (see Figures 2.c and 2.d). The two poles are closer by $\Delta\xi=0.02356$ compared to spontaneous emission. They describe the resonance and band width of the two stimulated emission spectrum peaks, with maxima of $W_1=14.36$ and $W_2=10.93$ (in units of ω_{ref}) (see Figure 3.a). Compared to spontaneous emission, the two maxima are reduced by $\Delta W_1=7.51$ and $\Delta W_2=4.73$. If an emission probability of almost zero is assumed after the second oscillation, then the stimulated emission decay time is $\tau=113.5$ (in units of $1/\omega_{ref}$) (see Figure 4.a). Compared to spontaneous emission, this time is increased by $\Delta\tau=19.2$. The phase difference of the two laser beams thus enables control of atomic decay time and of active cavity delay time [39-41]. At this point a less than ideal delay line has been designed, in which an input pulse is retarded and amplified but somewhat distorted. The plots of Figures 5.a and 5.b show that, compared to spontaneous emission, there is a narrower pass band, with $\xi = \omega-\Omega \approx (-0.04, 0.04)$, with the global transmission spectrum exhibiting similar values, so $G_{C,N+1} \in (G_{min}, G_{max}) = (3.270, 9.24)$ (in units of ω_{ref}), and most significantly the acceleration of coupling is now slightly modulated around the value $v_{C,N+1}$=0.05310 (in units of ω_{ref}/v_{ref}).

Finally, stimulated emission is considered in the presence of a degree of detuning, when the atom inside the symmetric QW 1D-PBG cavity remains coupled to the $(N+1)^{th}$ QNM, but no longer oscillates at the $(N+1)^{th}$ QNM frequency. The active delay line design can be improved with a final degree of freedom by varying the frequency detuning of the atom - $(N+1)^{th}$ QNM coupling (56). The application of maximum detuning is proposed to improve the active delay line. The atomic resonance Ω is lowered to within the photonic band gap, close to the $(N+1)^{th}$ QNM frequency $\mathrm{Re}\,\omega_{N+1}$, and the atom only remains coupled to the $(N+1)^{th}$ QNM if the atomic resonance is within the limit

$$\Omega = \mathrm{Re}\,\omega_{N+1} - \left|\mathrm{Im}\,\omega_{N+1}\right|, \tag{59}$$

when detuning is maximum:

$$\Delta_{N+1} = \frac{\mathrm{Re}\,\omega_{N+1} - \Omega}{R_{N+1}} = \frac{\left|\mathrm{Im}\,\omega_{N+1}\right|}{R_{N+1}}. \tag{60}$$

Detuned in this way, the two stimulated emission spectrum poles are shifted by the resonance Ω and exhibit real parts $\mathrm{Re}\xi_1$=0.03738 and $\mathrm{Re}\xi_2$=–0.07380, and imaginary parts $\mathrm{Im}\xi_1$=0.01176 and $\mathrm{Im}\xi_2$=0.02588 (both in units of ω_{ref}) (see Figures 2.c and 2.d). Compared to perfect tuning, the real parts are reduced by $\Delta\mathrm{Re}\xi_1$=0.01467 and $\Delta\mathrm{Re}\xi_2$=0.02175, while one imaginary part is reduced by $\Delta\mathrm{Im}\xi_1$=0.00611 and the other is raised by $\Delta\mathrm{Im}\xi_2$=0.0061. They describe the resonance and bandwidth of the two peaks of the stimulated emission spectrum when detuned, with maxima of W_1=38.83 and W_2=0.2974 (in units of ω_{ref}) (see Figure 3.b). Compared to perfect tuning, the first peak is raised by ΔW_1=24.47 and the second peak is lowered by ΔW_2=10.63. If the atomic emission probability is assumed to be almost zero after the second oscillation, then the stimulated emission decay time (linked to the active cavity delay time) in the detuned example is τ=111.2 (in units of $1/\omega_{ref}$) (see Figure 4.b). Compared to perfect tuning, the emission probability (and thus the input pulse) is somewhat warped and retarded by $\Delta\tau$=2.3.

The result is the design of a close to ideal delay line, with an input pulse being retarded, amplified and only slightly distorted. The plots of Figures 5.c and 5.d show that compared to stimulated emission, the detuned example has an even narrower pass band, at $\xi = \omega - \Omega \approx (0.02, 0.06)$, with the global transmission spectrum exhibiting higher values, with $G_{C,N+1} \in (G_{min}, G_{max}) = (6.005, 36.77)$ (in units of $1/\omega_{ref}$). Most significantly, the coupling acceleration is completely no modulated and almost constant at $v_{C,N+1} \cong 0.007182$ (in units of ω_{ref}/v_{ref}). As seen in Figure 5.d, the coupling acceleration modulation is shifted into the unused frequency range $\xi \approx (-0.07, -0.04)$.

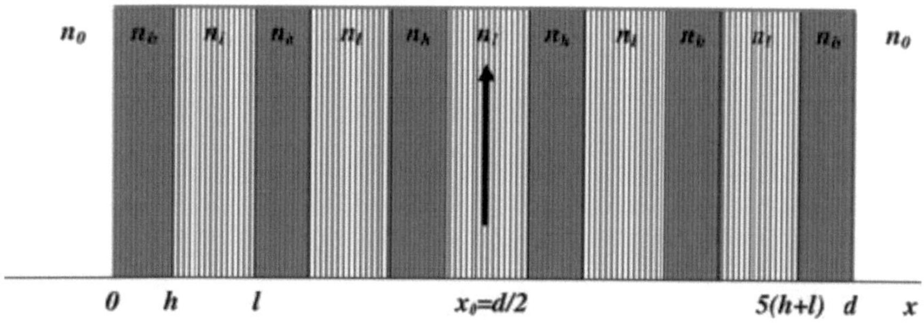

Figure 1. Symmetric Quarter-Wave (QW) one dimensional (1D) Photonic Band Gap (PBG) cavity with $\lambda_{ref}=1\mu m$ as reference wavelength, N=5 periods, consisting of two layers with refractive indices n_h=2 and n_l=1.5 and lengths $h=\lambda_{ref}/4n_h$ and $l=\lambda_{ref}/4n_l$. Terminal layers of the symmetric QW 1D-PBG cavity with parameters: n_h and $h=\lambda_{ref}/4n_h$. Length of the 1D-PBG cavity: $d=N(h+l)+h$. One atom is present, embedded in the centre of the 1D-PBG, so that $x_0=d/2$ (Figure reproduced from references [32,33]).

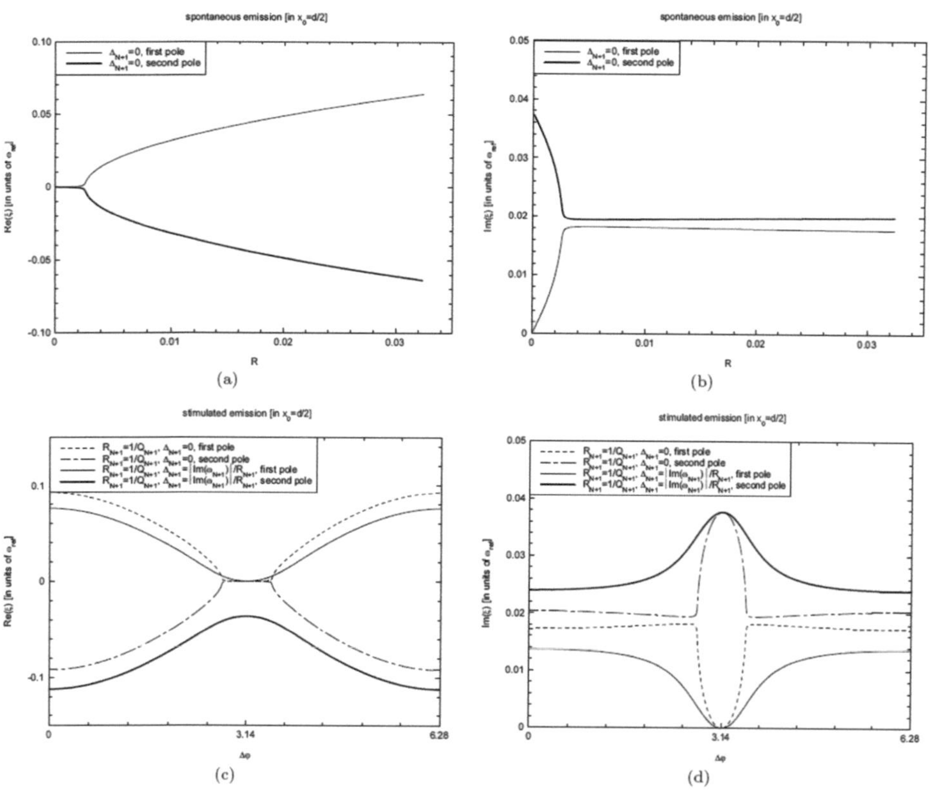

Figure 2. If the atom embedded inside the 1D-PBG cavity of Figure 1 oscillates at the $(N+1)^{th}$ Quasi-Normal Mode (QNM), close to the high-frequency band limit [i.e. perfectly tuned Δ_{N+1}=0, see Equation (56)], spontaneous emission under strong coupling regime exhibit two characteristic atomic emission spectrum poles, each pole being shifted by the atomic resonance Ω [see Equations (29) and (22)]; the real (Figure 2.a) and imaginary (Figure 2.b) parts, in units of the 1D-PBG reference frequency ω_{ref}, are plotted as functions of the degree of coupling $R={}_0/\Omega$, this being the ratio between the atomic decay-rate in vacuum $_0$, and resonance Ω [see Equation (55)].

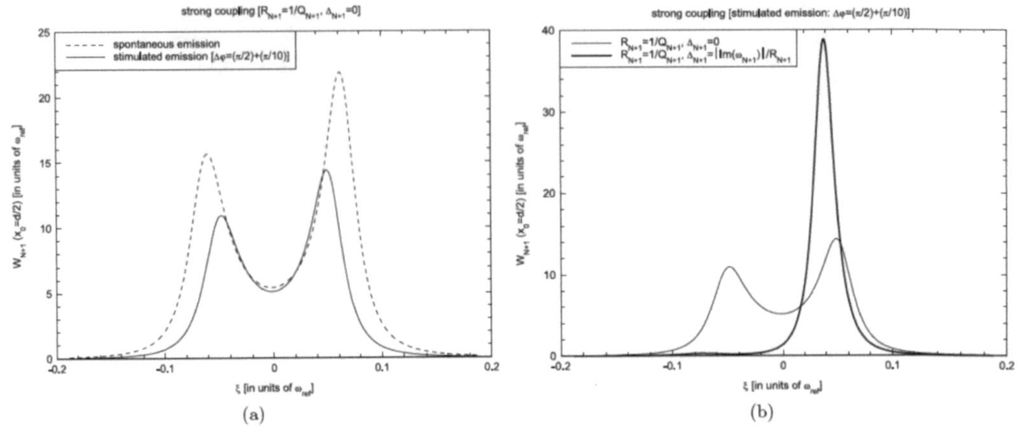

Figure 3. The emission spectrum of the atom embedded inside the 1D-PBG cavity of Figure 1 is plotted in units of the 1D-PBG reference frequency ω_{ref}, and as a function of the dimensionless shifted frequency $\xi=(\omega-\Omega)/\omega_{ref}$, with Ω denoting atomic resonance. The atom is coupled to the $(N+1)^{th}$ QNM frequency and emission occurs under strong coupling regime [for $R_{N+1}=1/Q_{N+1}$]. Figure 3.a illustrates hypothetical tuning, with the spontaneous atomic emission spectrum [see Equations (47) and (48)] compared to the stimulated emission spectrum [see Equations (47) and (48)], when the 1D-PBG is pumped by paired laser beams with an appropriate phase difference: $\Delta\varphi=(\pi/2)+(\pi/10)$ [see Equation (58)]. Figure 3.b instead illustrates a case of stimulated emission, comparing the perfectly tuned atomic emission spectrum with the detuned emission spectrum (Figure reproduced from references [32,33]).

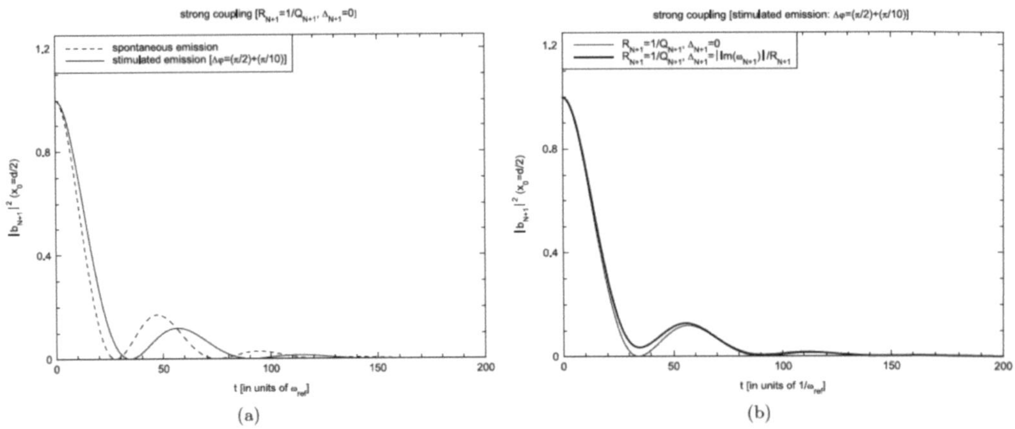

Figure 4. The emission probability of the atom embedded inside the 1D-PBG cavity of Figure 1 is plotted as a function of the normalized time $\omega_{ref}t$, with ω_{ref} being the 1D-PBG reference frequency. With reference to the operative conditions of Figure 3: hypothetical tuning is shown in Figure 4.a, comparing the spontaneous atomic emission probability [see Equation (34)], with stimulated emission probability [see Equation (35)] when the 1D-PBG is pumped by paired laser beams with an appropriate phase difference: $\Delta\varphi=(\pi/2)+(\pi/10)$. Figure 4.b illustrates stimulated emission, comparing atomic emission probability under perfect tuning with emission probability when detuned (Figure reproduced from references [32,33]).

If a pair of counter-propagating laser beams are tuned to the resonance Ω and the atom is coupled to the $(N+1)^{th}$ QNM, [i.e. $Q_{N+1}=\Omega\,/\mathrm{Im}\omega_{N+1}$, see Equation (54)], the stimulated emission

http://dx.doi.org/10.5772/59900

under strong coupling [for $R_{N+1}=1/Q_{N+1}$, see Equation (57)] exhibits two characteristic atomic emission spectrum poles, each pole being shifted by the resonance Ω [see Equations (29) and (26)]. The real (Figure 2.c) and imaginary (Figure 2.d) parts, in units of the 1D-PBG reference frequency ω_{ref}, are plotted as functions of the phase difference $\Delta\varphi$ between the paired laser beams regardless of whether the atom oscillates at the $(N+1)^{th}$ QNM frequency [i.e. perfectly tuned $\Delta_{N+1}=0$] or at a frequency in the band gap close to the high frequency band limit [i.e. detuned $\Delta_{N+1}=\mathrm{Im}\omega_{N+1}/R_{N+1}$, see Equation (60)] (Figure reproduced from references [32,33]).

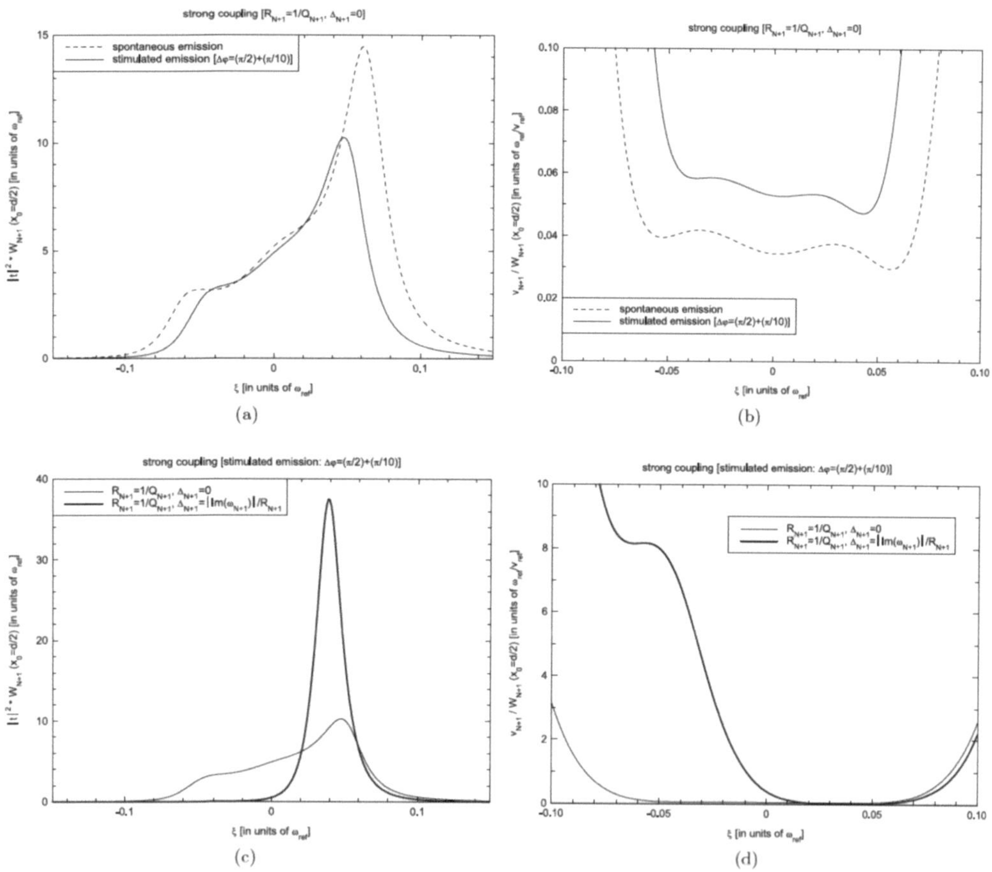

Figure 5. A delay line using an active cavity comprising the 1D-PBG cavity plus atom (Figure 1) can be designed by characterizing the line according to global transmission [see Equation (52)], and "coupling acceleration" [see Equation (53)] of the electromagnetic (e.m.) field. Global transmission, in units of the 1D-PBG reference frequency ω_{ref}, and coupling acceleration, in units of ω_{ref}/v_{ref}, being v_{ref} the group velocity of the e.m. field in vacuum, are plotted as functions of the dimensionless shifted frequency $\xi=(\omega-\Omega)/\omega_{ref}$, with Ω denoting atomic resonance. With reference to the operative conditions of Figure 3: perfect tuning is shown in Figure 5.a (Figure 5.b), comparing the global transmission (coupling acceleration) of the active delay line for spontaneous emission, with the global transmission (coupling acceleration) for stimulated emission when the 1D-PBG is pumped by paired laser beams with an appropriate phase difference: $\Delta\varphi=(\pi/2)+(\pi/10)$. Figure 5.c (Figure 5.d) compares the global transmission (coupling acceleration) of the active delay line when perfectly tuned, with the global transmission (coupling acceleration) when detuned under stimulated emission (Figure reproduced from references [32,33]).

8. Final discussion and concluding remarks

This chapter discussed atomic stimulated emission processes, under strong coupling, inside a one dimensional (1D) Photonic Band Gap (PBG) cavity, which is pumped by a pair of counter-propagating laser beams [32,33]. The atom-field interaction was modelled by quantum electro-dynamics, with the atom considered as a two level system, the electromagnetic (e.m.) field as superposition of its normal modes, and applying the dipole approximation, the Wigner-Weisskopf equations of motion, and the rotating wave approximations. The unenclosed cavity example under investigation was approached applying the Quasi-Normal Mode (QNM), while the local density of states (LDOS) was interpreted as the local probability density of exciting a single QNM within the cavity. In this approach, the LDOS depends on the phase difference of the paired laser beams, and the most significant result is that the strong coupling regime can occur with high LDOS values. The investigation also confirms the well known phenomenon [39-41] that atomic emission probability decays with oscillation, causing the atomic emission spectrum to split into two peaks (Rabi splitting). The novelty that emerged in this chapter is that it appears to be possible to coherently control both the atomic emission probability oscillations and the Rabi splitting of the emission spectrum using the phase difference of the paired laser beams. Finally, some criteria were proposed for the design of an active cavity comprising a 1D-PBG cavity plus atom, to serve as an active delay line. It is seen that suitable phase differences between the paired laser beams make it possible to achieve high delayed pulse transmission in a narrow pass band.

The issue of e.m. field interaction with atoms when the e.m. modes are conditioned by the environment (inside a cavity, or proximal to walls) can be approached in several ways. For example, the dynamics of the e.m. field can first be established inside and outside the cavity (or proximal/distant from walls), and then atomic coupling with the normal modes (NMs) of the combined system [42-46] can be considered. An alternative approach applies the discrete (dissipative) QNMs of the unenclosed cavity in place of the continuous (Hermitian) NMs. When applying the QNMs, the internal field cavity is coupled to the external e.m. fields (beyond the two cavity limits) by boundary conditions [47-50].

A third approach is proposed in the present chapter, combining both those described above with the aim of merging their analytic potentials. Canonical quantum electro-dynamics is applied for the definition of an e.m field as a superposition of NMs, while an unenclosed cavity is defined adopting a QNM approach, when LDOS is interpreted as the local probability density of exciting a single QNM of the cavity. The DOS is linked to the cavity boundary conditions. The e.m. field satisfies incoming and outgoing wave conditions on the cavity surfaces, and so the DOS depends on the externally pumped photon reservoir. When the cavity is excited by paired counter-propagating pumps, the DOS expresses the probability distribution of exciting a single QNM of the cavity.

In the case of spontaneous emission, the paired pumps are modelled as vacuum fluctuations from the ground state of the e.m. field, while the DOS is construed simply as a feature of cavity geometry. Instead, in the case of stimulated emission, the paired pumps are modelled as two laser beams in a coherent state, so that the DOS depends on the cavity geometry and can be controlled by the phase difference of the paired laser beams. These results clearly highlight how the DOS of an unenclosed cavity is determined by the cavity excitation state.

Acknowledgements

The author, Dr. Alessandro Settimi, is extremely grateful to Dr. Sergio Severini for his outstanding support and friendship, to Prof. Concita Sibilia and Prof. Mario Bertolotti for their interesting pointers to literature regarding photonic crystals, and to Prof. Anna Napoli and Prof. Antonino Messina for their valuable discussions regarding stimulated emission.

Author details

Alessandro Settimi*

Address all correspondence to: alessandro.settimi@ingv.it

Istituto Nazionale di Geofisica e Vulcanologia (INGV), Sezione di Geomagnetismo, Aeronomia e Geofisica Ambientale (ROMA 2), Rome, Italy

References

[1] Purcell E. M. Spontaneous emission probabilities at radio frequencies. Physical Review 1946; 69, 681.

[2] Kleppner D. Inhibited Spontaneous Emission. Physical Review Letters 1981; 47 (4) 233-236, DOI: /10.1103/PhysRevLett.47.233.

[3] Drexhage K. H. Interaction of light with monomolecular dye layers. In: Wolf E. (ed.) Progress in Optics. New York: North-Holland; 1974. vol. 12, p. 165.

[4] Goy P., Raimond J. M., Gross M., Haroche S. Observation of Cavity-Enhanced Single-Atom Spontaneous Emission. Physical Review Letters 1983; 50 (24) 1903-1906, DOI: 10.1103/PhysRevLett.50.1903.

[5] Hulet R. G., Hilfer E. S., Kleppner D. Inhibited Spontaneous Emission by a Rydberg Atom. Physical Review Letters 1985; 55 (20) 2137-2140, DOI: 10.1103/PhysRevLett.55.2137.

[6] Jhe W., Anderson A., Hinds E. A., Meschede D., Moi L., Haroche S. Suppression of spontaneous decay at optical frequencies: Test of vacuum-field anisotropy in confined space. Physical Review Letters 1987; 58 (7) 666-669, DOI: 10.1103/PhysRevLett.58.666.

[7] Heinzen D. J., Childs J. J., Thomas J. E., Feld M. S. Enhanced and inhibited visible spontaneous emission by atoms in a confocal resonator. Physical Review Letters 1987; 58 (13) 1320-1323, DOI: 10.1103/PhysRevLett.58.1320.

[8] De Martini F., Innocenti G., Jacobovitz G. R., Mataloni P. Anomalous Spontaneous Emission Time in a Microscopic Optical Cavity. Physical Review Letters 1987; 59 (26) 2955-2928, DOI: 10.1103/PhysRevLett.59.2955.

[9] Jaynes E. T., Cummings F. W. Comparison of quantum and semiclassical radiation theories with application to the beam maser. Proceedings of the IEEE 1963; 51 (1) 89-109, DOI: 10.1109/PROC.1963.1664.

[10] Rempe G., Walther H., Klein N. Observation of quantum collapse and revival in a one-atom maser. Physical Review Letters 1987; 58 (4) 353-356, DOI: 10.1103/PhysRevLett.58.353.

[11] John S. Electromagnetic Absorption in a Disordered Medium near a Photon Mobility Edge. Physical Review Letters 1984; 53 (22) 2169-2172, DOI: 10.1103/PhysRevLett.53.2169.

[12] Yablonovitch E. Inhibited Spontaneous Emission in Solid-State Physics and Electronics. Physical Review Letters 1987; 58 (20) 2059-2062, DOI: 10.1103/PhysRevLett.58.2059.

[13] John S. Strong localization of photons in certain disordered dielectric superlattices. Physical Review Letters 1987; 58 (23) 2486-2489, DOI: 10.1103/PhysRevLett.58.2486.

[14] John S., Wang J. Quantum electrodynamics near a photonic band gap: Photon bound states and dressed atoms. Physical Review Letters 1990; 64 (20) 2418-2421, DOI: 10.1103/PhysRevLett.64.2418.

[15] John S., Wang J. Quantum optics of localized light in a photonic band gap. Physical Review B 1991; 43 (16) 12772-12789, DOI: 10.1103/PhysRevB.43.12772.

[16] Burstein E., Weisbuch C., editors. Confined Electrons and Photons: New Physics and Applications. New York: Plenum Press; 1995. p. 907.

[17] Nabiev R. F., Yeh P., Sanchez-Mondragon J. J. Dynamics of the spontaneous emission of an atom into the photon-density-of-states gap: Solvable quantum-electrodynamical model. Physical Review A 1993; 47 (4) 3380-3384, DOI: 10.1103/PhysRevA.47.3380.

[18] Lai H. M., Leung P. T., Young K. Electromagnetic decay into a narrow resonance in an optical cavity. Physical Review A 1988; 37 (5) 1597-1606, DOI: 10.1103/PhysRevA.37.1597.

[19] Leung P. T., Liu S. Y., Young K. Completeness and orthogonality of quasinormal modes in leaky optical cavities. Physical Review A 1994; 49 (4) 3057-3067, DOI: 10.1103/PhysRevA.49.3057.

[20] Leung P. T., Tong S. S., Young K. Two-component eigenfunction expansion for open systems described by the wave equation I: completeness of expansion. Journal of

Physics A: Mathematical and General 1997; 30 (6) 2139-2151, DOI: 10.1088/0305-4470/30/6/034.

[21] Leung P. T., Tong S. S., Young K. Two-component eigenfunction expansion for open systems described by the wave equation II: linear space structure. Journal of Physics A: Mathematical and General 1997; 30 (6) 2153-2162, DOI: 10.1088/0305-4470/30/6/035.

[22] Ching E. S. C., Leung P. T., Maassen van der Brink A., Suen W. M., Tong S. S., Young K. Quasinormal-mode expansion for waves in open systems. Reviews of Modern Physics 1998; 70 (4) 1545-1554, DOI: 10.1103/RevModPhys.70.1545.

[23] Settimi A., Severini S., Mattiucci N., Sibilia C., Centini M., D'Aguanno G., Bertolotti M., Scalora M., Bloemer M., Bowden C. M. Quasinormal-mode description of waves in one-dimensional photonic crystals. Physical Review E 2003; 68 (2) 026614 [1-11], DOI: 10.1103/PhysRevE.68.026614.

[24] Severini S., Settimi A., Sibilia C., Bertolotti M., Napoli A., Messina A. Quasi-Normal Frequencies in Open Cavities: An Application to Photonic Crystals. Acta Physica Hungarica B 2005; 23 (3-4) 135-142, DOI: 10.1556/APH.23.2005.3-4.3.

[25] Cohen-Tannoudji C., Diu B., Laloe F. Quantum Mechanics. New York: John Wiley & Sons Inc., Second Volume Set Edition; 1977. p. 1524.

[26] Louisell W. H. Quantum Statistical Properties of Radiation (Pure & Applied Optics). New York: John Wiley & Sons Inc., First Edition; 1973. p. 544.

[27] Ho K. C., Leung P. T., Maassen van den Brink A., Young K. Second quantization of open systems using quasinormal modes. Physical Review E 1998; 58 (3) 2965-2978, DOI: 10.1103/PhysRevE.58.2965.

[28] Severini S., Settimi A., Sibilia C., Bertolotti M., Napoli A., Messina A. Second quantization and atomic spontaneous emission inside one-dimensional photonic crystals via a quasinormal-modes approach. Physical Review E 2004; 70 (5) 056614, [1-12], DOI: 10.1103/PhysRevE.70.056614.

[29] Severini S., Settimi A., Sibilia C., Bertolotti M., Napoli A., Messina A. Coherent control of stimulated emission inside one dimensional photonic crystals: strong coupling regime. European Physical Journal B 2006; 50 (3) 379-391, DOI: 10.1140/epjb/e2006-00165-2.

[30] Severini S., Settimi A., Sibilia C., Bertolotti M., Napoli A., Messina A. Erratum - Coherent control of stimulated emission inside one dimensional photonic crystals: strong coupling regime. European Physical Journal B 2009; 69 (4) 613-614, DOI: 10.1140/epjb/e2009-00201-9.

[31] Settimi A., Severini S., Sibilia C., Bertolotti M., Centini M., Napoli A., Messina A.. Coherent control of stimulated emission inside one-dimensional photonic crystals. Physical Review E 2005; 71 (6) 066606 [1-10], DOI: 10.1103/PhysRevE.71.066606.

[32] Settimi A., Severini S., Sibilia C., Bertolotti M., Napoli A., Messina A.. Coherent control of stimulated emission inside one dimensional photonic crystals: strong coupling regime. European Physical Journal B 2006; 50 (3) 379-391, DOI: 10.1140/epjb/e2006-00165-2.

[33] Settimi A., Severini S., Sibilia C., Bertolotti M., Napoli A., Messina A.. ERRATUM Coherent control of stimulated emission inside one dimensional photonic crystals: strong coupling regime. European Physical Journal B 2009; 69 (4) 613-614, DOI: 10.1140/epjb/e2009-00201-9.

[34] Abrikosov A. A., Gor'kov L. P. Dzyaloshinski I. E. Methods of Quantum Field Theory in Statistical Physics (Dover Books on Physics). New York: Dover Publications, Revised Edition; 1975. p. 384.

[35] Leung P. T., Maassen van den Brink A., Young K. Quasinormal-mode Quantization of Open Systems. In: Lim S. C., Abd-Shukor R., Kwek K. H. (eds) Frontiers in Quantum Physics: proceedings of the International Conference, July 1998, University Kelangsoon, Singapore, Malaysia. Singapore: Springer-Verlag; 1998. p. 214-218.

[36] Blow K. J., Loudon R., Phoenix S. J. D., Shepherd T. J. Continuum fields in quantum optics. Physical Review A 1990; 42 (7) 4102-4114, DOI: 10.1103/PhysRevA.42.4102.

[37] Carrier G. F., Krook M., Pearson C. E. Functions of a complex variable – theory and technique. New York: McGraw-Hill Book Company, First Edition; 1966. p. 438.

[38] Hughes S., Kamada H. Single-quantum-dot strong coupling in a semiconductor photonic crystal nanocavity side coupled to a waveguide. Physical Review B 2004; 70 (19) 195313 [1-5], DOI: 10.1103/PhysRevB.70.195313.

[39] Allen L., Eberly, J. H. Optical Resonance and Two-Level Atoms (Dover Books on Physics). New York: Dover Pubblication; 1987. p. 256.

[40] Bonifacio R., Lugiato L. A. Cooperative radiation processes in two-level systems: Superfluorescence. Physical Review A 1975; 11 (5) 1507-1521, DOI: 10.1103/PhysRevA.11.1507.

[41] Sakoda K., Haus J. W. Superfluorescence in photonic crystals with pencil-like excitation. Physical Review A 2003; 68 (5) 053809 [1-5], DOI: 10.1103/PhysRevA.68.053809.

[42] Louisell W. H., Walker L. R. Density-Operator Theory of Harmonic Oscillator Relaxation. Physical Review 1965; 137 (1B) B204-B211, DOI: 10.1103/PhysRev.137.B204.

[43] Senitzky I. R. Dissipation in Quantum Mechanics. The Harmonic Oscillator. Physical Review 1960; 119 (2) 670-679, DOI: 10.1103/PhysRev.119.670.

[44] Senitzky I. R. Dissipation in Quantum Mechanics. The Harmonic Oscillator. II. Physical Review 1961; 124 (3) 642-648, DOI: 10.1103/PhysRev.124.642.

[45] Senitzky I. R. Dissipation in Quantum Mechanics. The Two-Level System. Physical Review 1963; 131 (6) 2827-2838, DOI: 10.1103/PhysRev.131.2827.

[46] Lax M. Quantum Noise. IV. Quantum Theory of Noise Sources. Physical Review 1966; 145 (1) 110-129, DOI: 10.1103/PhysRev.145.110.

[47] Dekker H. A note on the exact solution of the dynamics of an oscillator coupled to a finitely extended one-dimensional mechanical field and the ensuing quantum mechanical ultraviolet divergence. Physics Letters A 1984; 104 (2) 72-76, DOI: 10.1016/0375-9601(84)90965-4.

[48] Dekker H. Particles on a string: Towards understanding a quantum mechanical divergence. Physics Letters A 1984; 105 (8) 395-400, DOI: 10.1016/0375-9601(84)90715-1.

[49] Dekker H. Bound electron dynamics: Exact solution for a one-dimensional oscillator-string model. Physics Letters A 1984; 105 (8) 401-406, DOI: 10.1016/0375-9601(84)90716-3.

[50] Dekker H. Exactly solvable model of a particle interacting with a field: The origin of a quantum-mechanical divergence. Physical Review A 1985; 31 (2) 1067-1076, DOI: 10.1103/PhysRevA.31.1067.